Genetics
of Domestic Animals

Genetics of Domestic Animals

CHARLES E. STUFFLEBEAM

Southwest Missouri State University

PRENTICE HALL, Englewood Cliffs, New Jersey 07632

Library of Congress Cataloging-in-Publication Data

Stufflebeam, Charles E.
    Genetics of domestic animals / Charles E. Stufflebeam.
        p.   cm.
    Includes index.
    ISBN 0-13-351214-2
    1. Domestic animals--Genetics.   2. Animal breeding.   I. Title.
SF105.S92 1989
636.08'21--dc19   88-27419
                                                                  CIP

Editorial/production supervision: Merrill Peterson
Interior design: Joan L. Stone
Cover design: Diane Saxe
Manufacturing buyer: Bob Anderson

© 1989 by Prentice-Hall, Inc.
A Division of Simon & Schuster
Englewood Cliffs, New Jersey 07632

Printed in the United States of America

10   9   8   7   6   5   4   3   2   1

ISBN 0-13-351214-2

Prentice-Hall International (UK) Limited, *London*
Prentice-Hall of Australia Pty. Limited, *Sydney*
Prentice-Hall Canada Inc., *Toronto*
Prentice-Hall Hispanoamericana, S.A., *Mexico*
Prentice-Hall of India Private Limited, *New Delhi*
Prentice-Hall of Japan, Inc., *Tokyo*
Simon & Schuster Asia Pte. Ltd., *Singapore*
Editora Prentice-Hall do Brasil, Ltda., *Rio de Janeiro*

Contents

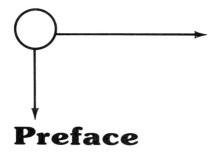

Preface

The materials in this book have been designed for second and third year college and university students with an interest in domestic animals. It should be well suited for students in agriculture, biology, and veterinary medicine. Many excellent genetics textbooks are available, but very few of them use domestic animals as their primary models.

The book is designed so that only a minimal level of preparation in biology, chemistry, and mathematics is needed, most of which could be obtained at the secondary level or first year in college. A knowledge and understanding of the contents of this book should prepare a student very well for an advanced undergraduate or graduate course in genetics or animal breeding.

Each chapter begins with an outline of the material presented, which should be helpful in previewing the chapter, as well as serve as a guide for later review. Study questions, exercises, and problems follow each chapter to assist and guide students and instructors in the study of the principles and concepts presented. A list of other reading sources is also included at the end of each chapter.

The book is divided into three distinct sections. The first part deals with the biology of the cell and chemistry of the gene. The second part, chapters two through nine, covers qualitative genetics—basic principles of genetics. Topics included here are one and two pair crosses, multiple alleles, sex-related inheritance, linkage, epistasis, and probabilities. The third part deals with "quantitative genetics"—selection, inbreeding, and outbreeding.

PART 1
BIOLOGY AND CHEMISTRY OF INHERITANCE

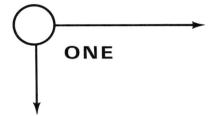

ONE

Physiological Basis of Inheritance

Genetics is a biological science that deals with heredity and variation. *Heredity* involves the transmission of genetic material (genes) from one generation to the next. *Variation* includes the differences that can be seen among characteristics within each species. Differences among animal characteristics are brought about, in part, by the transmission of changed genes (mutations) or new combinations of existing genes. Environmental factors also play a significant role as a source of variation. The basic pattern of a trait or characteristic is determined by heredity; the development of the trait is then influenced in varying degrees by environment, as well as by other genes.

Genetics as a science began in the 1860s with the work of Gregor Mendel, an Austrian monk, who discovered the basic law of segregation and recombination of genes. He did not know about genes as such, but worked out the one-pair and two-pair ratios which are basic to the study of genetics. The results of Mendel's work with garden peas were published in 1865, but remained in relative obscurity for nearly 35 years. In 1900, three other scientists, working independently of each other, rediscovered Mendel's principles and then found the results of his work.

1.1 CELLS AND CHROMOSOMES

The smallest structural unit of an animal's body is the *cell* (Figure 1.1). A typical cell is microscopic, varying in size from approximately 10 to 20 micrometers in diameter (1 micrometer equals approximately $\frac{1}{25}$ inch). The

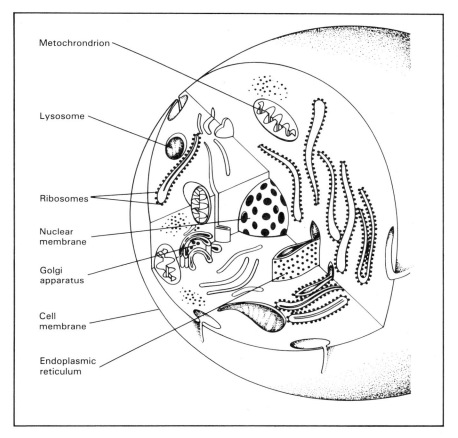

Figure 1.1 Sections and parts of a typical animal cell.

outer covering of an animal cell is the *cell membrane*. Inside the cell membrane are the *cytoplasm* and the *nucleus*. Contained within the cytoplasm are a number of *organelles* that are associated with the many vital functions of the cell. Included among these organelles are the mitochondria, the endoplasmic reticulum, lysosomes, ribosomes, and Golgi bodies. The nucleus is covered with a membrane inside of which are found the chromosomes and genes. It is with the genes and chromosomes that the subject of genetics is most concerned.

1.1.1 Characteristics of Chromosomes

Chromosomes are relatively slender, threadlike strands of material that contain the units of inheritance called *genes*. One of the prominent features of most chromosomes is the centromere. Although the word *centromere* means "central part," it is not always located in the center of the chromosome. The position of the centromere and the length of the chromosome are two major factors involved in distinguishing one chromosome from another.

Chromosomes may be categorized into three or four groups based on the position of the centromere. Figure 1.2 is a diagram of four chromosomes differing only by the position of their centromeres. A chromosome with its centromere located approximately in the center is called a *metacentric* chromosome. The two arms of a metacentric chromosome will be very nearly the same size. If the centromere is located far enough from the center that the two arms of the chromosome are distinctly unequal in size, the term *submetacentric* is used. An *acrocentric* chromosome is one with the centromere located very near the end. If the centromere is located at the very end of a chromosome, the term *telocentric* is used. Sometimes it is very difficult to distinguish between a telocentric and an acrocentric chromosome.

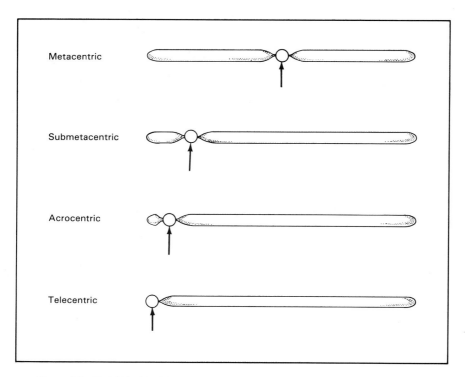

Figure 1.2 Four kinds of chromosomes categorized according to the positions of their centromeres.

The overall length of chromosomes varies from a few to several micrometers. The number of genes may vary from a few on the very short chromosomes, to possibly thousands on the longer ones.

1.1.2 Chromosome Number

The number of chromosomes in the nuclei of body cells is constant among animals of the same species. All body cells of cattle, for example, would be expected to have 60 chromosomes. If for some reason a zygote is formed that is missing one or more chromosomes, the genes that are found on those chromosomes will also be missing. Some of the missing genes could very well control functions so vital to the animal that it would not be able to survive without them. The presence of extra chromosomes results in the presence of extra genes, a situation that can cause metabolic problems for the animal. In fact, the chromosome number is so important that deviations from the normal affect the animal to the extent that it will usually die during embryonic development. The typical number of chromosomes in the body cells of several species is listed in Table 1.1.

In addition to the fact that there is a constant number of chromosomes in the body cells of animals, the chromosomes exist in pairs. The members of each of these pairs are of the same shape and length and have their centromeres located in the same position. The sex chromosomes represent an exception to this statement. The Y chromosome in male mammals is generally smaller than the X chromosome and does not carry as many genes. The situation regarding the Z and W chromosomes in female birds is similar. A more complete explanation of the characteristics and functions of the sex chromosomes is presented in Chapter Five.

Since the chromosomes in body cells do exist in pairs, it is customary to express chromosome number in terms of the number of pairs typical for the species, such as 19 pairs for swine, 39 pairs for dogs, and 23 pairs for humans. The two members of a pair are called *homologs* of each other. Homologous chromosomes are defined as two chromosomes that are alike

**TABLE 1.1   Chromosome Numbers in Diploid Cells
of Several Animal Species**

Pig	38	Bison	60
Cat	38	Cattle	60
Mouse	40	Goat	60
Rat	42	Donkey	62
Rabbit	44	Horse	64
Human	46	Dog	78
Sheep	54	Chicken	78

in size, shape, and position of their centromeres. Homologous chromosomes may or may not contain some of the same genes.

Cells that contain pairs of chromosomes are described as being *diploid*. The symbol for diploidy is 2*n*. *Body cells*, also called *somatic cells*, are usually diploid. *Sex cells*, or *gametes*, are generally haploid (*n*). In gametes, the chromosomes are not paired. Only one chromosome from each of the homologous pairs is generally found in the nucleus of a gamete.

Figure 1.3 is a photograph of the chromosomes of a boar, taken during the metaphase of mitosis. The individual chromosomes have been cut out of the photograph and arranged by homologous pairs according to their overall length. The pairs of autosomes were then numbered from the largest to the smallest, with the sex chromosomes being placed last in the arrangement. Autosomes include all chromosomes except the sex chromosomes. Such an arrangement of chromosomes is called a *karyotype*. Karyotypes can be useful in determining chromosome number; they may be used to help

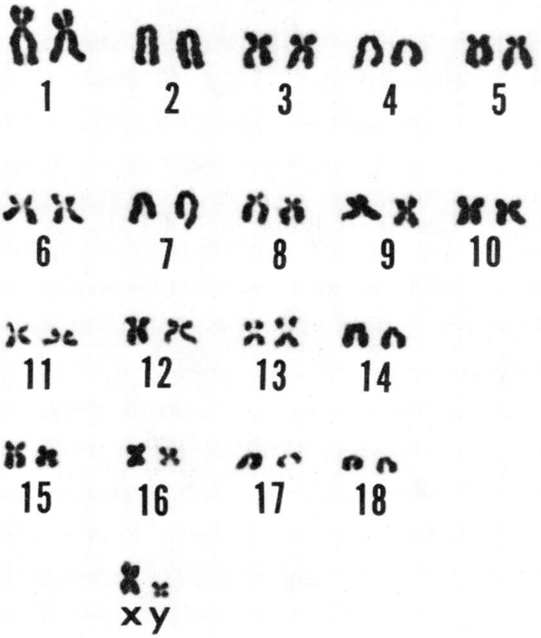

Figure 1.3 Photomicrograph of chromosomes of the pig arranged in a karyotype, with the chromosomes in pairs. At the metaphase of cell division when the chromosomes can be stained and observed, they appear doubled (sister chromatids) and are connected at the centomere. [From John F. Lasley, *Genetics of Livestock Improvement*, 4/E, © 1987, p. 19. Reprinted by permission of Prentice-Hall, Inc., Englewood Cliffs, NJ.]

determine the sex of embryos; and deviations from normal morphology and chromosome number can be identified by studying karyotypes.

1.1.3 The Concept of Alleles

The position or location on a chromosome where a particular gene is found is called a *locus*. Every gene has a locus. A pair of homologous chromosomes may or may not possess identical genes at the same locus. However, even when the genes at two homologous loci are not identical, they will affect the same trait. The different forms of the genes that can be found at the same loci on homologous chromosomes are called *alleles*. In summary, three key points characterize the definition of alleles: the genes are different; they occupy the same loci (positions) on homologous chromosomes; and they have their effects on the same trait. For example, a particular gene may cause a cow to develop horns, while its allele may cause another cow to be polled. The presence or absence of horns is the trait for which there are two alternate phenotypes or expressions: horns and polled. The term *phenotype* refers to the way in which genes express themselves.

When describing genes or solving genetics problems, it is necessary to use symbols to represent the genes. The most common practice is to use letters of the alphabet. There is no absolute standard to follow, but the most common way is to use upper- and lowercase letters to represent the alleles if only two are involved. Suppose that at one locus two alleles are possible. We can refer to this locus as the A-locus and let the letters A and a be the symbols for the two alleles. Either allele can occupy the locus on either homologous chromosome. In a group of animals it is very likely that all possible combinations of the two genes will exist. Some animals might possess two of the A genes, in which case the genotype would be written as AA. Other animals may possess two of the a genes and be designated as aa. Since each of these genotypes is made up of identical genes, they are referred to as *homozygous* (*homo* means "the same"). A third genotype is possible, the one that contains one of each of the two alleles (Aa). Since the genotype is made up of genes that are not identical, it is referred to as *heterozygous* (*hetero* means "different").

1.2 THE CHEMICAL NATURE OF GENES AND CHROMOSOMES

1.2.1 The Structure of DNA

A gene is composed of a substance known as *deoxyribonucleic acid* (DNA). DNA consists of two relatively long strands of material twisted to form a helix or spiral-like structure (Figure 1.4). The chromosomes are

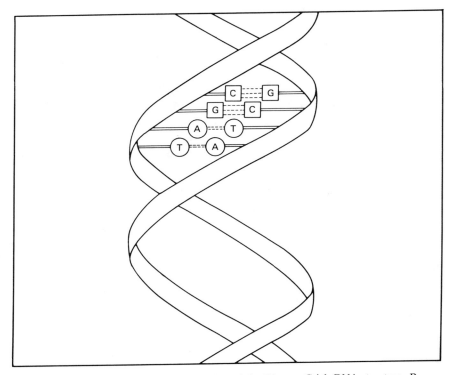

Figure 1.4　Diagrammatic representation of the Watson–Crick DNA structure. P, Phosphate; S, sugar; A, adenine; T, thymine; G, guanine; C, cytosine. The horizontal parallel lines symbolize hydrogen bonding between complementary bases.

made up of a complex substance called *nucleoprotein*, which consists of the DNA strands and a special kind of protein. Each of the two strands of DNA is composed of thousands of units called *nucleotides*. Each nucleotide is composed of a special nitrogenous base, the sugar deoxyribose, and phosphoric acid. The linkage in each strand of DNA is formed by a chemical connection between the sugar (deoxyribose) of one nucleotide and the phosphoric acid of the next, as illustrated in Figure 1.6.

　　The two strands of DNA are held together by relatively weak connections between the nitrogenous bases of one strand and those of the other strand. These relatively weak chemical connections are composed of hydrogen bonding. Four kinds of nitrogenous bases are found in DNA: adenine, thymine, cytosine, and guanine. The nature of the hydrogen bonding is such that adenine and thymine are attracted to each other, as are cytosine to guanine. This phenomenon is referred to as the *base-pairing* principle. (Figure 1.5)

Prior to cell division, actually, as a part of the entire process of cell division, each chromosome manufactures a new chromosome identical to itself. The first step in this duplication process involves the untwisting and separation of the two strands in the DNA helix. Each original strand then manufactures a new strand by attracting the appropriate nucleotides as determined by the base-pairing principle. This process is illustrated in Figure 1.6.

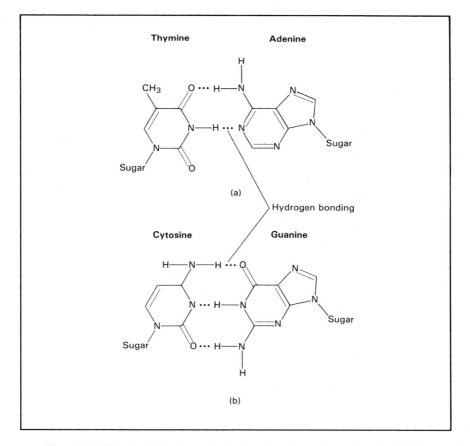

Figure 1.5   Pairings of thymine and adenine (a), and cytosine and guanine (b), by means of hydrogen bonding as in DNA.

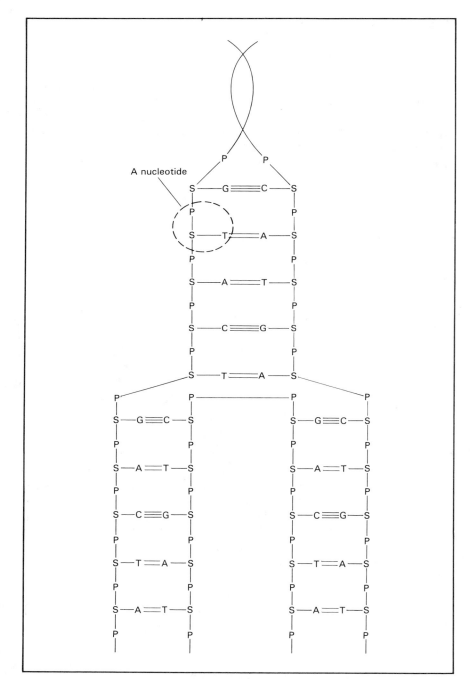

Figure 1.6 Illustration of a double-stranded DNA molecule duplicating itself. The two strands split and assemble the various parts on each strand, resulting in exact

1.2.2 Genetic Control of Protein Synthesis

One way that genes function is by controlling the production of enzymes that act as catalysts in various chemical reactions that take place in the cell, usually within the cytoplasm. Enzymes are proteins that are composed of amino acid subunits. Each enzyme is made up of a particular number and sequence of amino acids which is different from that of any other protein. The sequence of amino acids in each enzyme is determined by the sequence of nucleotides in the "coding" strand of DNA in the gene. A group of three nucleotides in the DNA is necessary to code a particular amino acid in the enzyme.

In order for the information carried in the genes to be transferred to the site of protein synthesis in the cytoplasm, a second kind of nucleic acid, called *ribonucleic acid* (RNA), comes into the picture. RNA is made up of the sugar ribose, phosphoric acid, and the four nitrogenous bases adenine, cytosine, guanine, and uracil. Uracil functions in RNA in much the same way as thymine does in DNA. In the process of base pairing, uracil pairs with thymine. The nucleotides in RNA are called *ribonculeotides*, while those of DNA are called *deoxyribonucleotides*. In summary, RNA differs from DNA in at least two ways: one is that RNA contains the sugar ribose instead of deoxyribose; the other is that RNA contains the base uracil in place of thymine.

One kind of RNA, called *messenger RNA*, or mRNA, is synthesized in the nucleus by the coding strand of DNA. One strand of DNA in a gene acts as a template in the synthesis of a molecule of messenger RNA. The mRNA, which consists of a single strand of nucleotides, contains the same number of nucleotides as are found in one strand of DNA in the gene from which it is patterned. Whenever the deoxyribonucleotides containing adenine, cytosine, guanine, and thymine appear in the template strand of DNA, the ribonucleotides containing uracil, guanine, cytosine, and adenine, respectively, are positioned in the molecule of mRNA. This is illustrated in Figure 1.7.

When the synthesis of a molecule of mRNA has been completed, it moves out of the nucleus and into the cytoplasm to form a complex with a ribosome. Ribosomes are relatively large structures usually associated with a system of membranes in the cytoplasm called the *endoplasmic reticulum.* The ribosomes are composed in part of ribonucleic acid, designated as *ribo-*

duplication of the molecule. P, phosphoric acid; S, sugar dioxyribose; A, base adenine; T, base thymine; G, base guanine; and C, base cytosine. Also note that a nucleotide containing P, S, and T is illustrated. [From John F. Lasley, *Genetics of Livestock Improvement,* 4/E, © 1987, p. 45. Reprinted by permission of Prentice-Hall, Inc., Englewood Cliffs, NJ.]

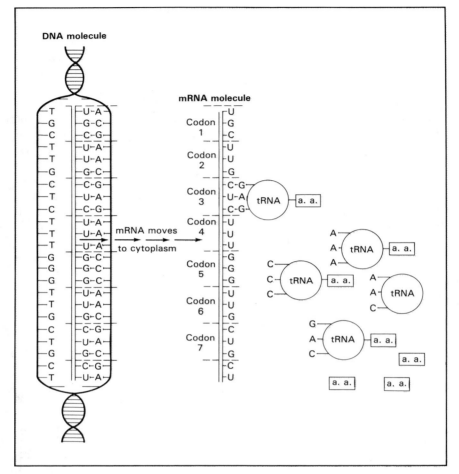

Figure 1.7 Illustration of how proteins are formed in the cytoplasm of the cell. mRNA is built in the nucleus by DNA and carries the code for protein formation to the ribosomes in the cytoplasm. The code is sent by codons (three consecutive bases), each of which specifies a particular amino acid. tRNA identifies a particular amino acid by means of an anticodon corresponding to the codon of the mRNA. tRNA carries this amino acid to the ribosome, where proteins are built along the mRNA molecule. The various amino acids are bound together to form molecules in the ribosomes according to the code sent by the means of mRNA. The kind of protein molecule built depends on the kind, number, and arrangement of amino acids in the protein molecule. [From John F. Lasley, *Genetics of Livestock Improvement,* 4/E, p. 46. Reprinted by permission of Prentice-Hall, Inc., Englewood Cliffs, NJ.]

somal RNA, or rRNA. They function as the site where protein molecules are synthesized.

Each set of three nucleotides in a molecule of messenger RNA serves to help position a particular amino acid in the enzyme, or other protein, being synthesized. Each set of three nucleotides is referred to as a *triplet* or *codon*. The sequence of triplets, or codons, in mRNA ultimately determines the sequence of amino acids in the protein being synthesized on a particular ribosome. This leads to a discussion of a third kind of RNA, called *transfer RNA*, or tRNA.

Transfer RNA is found in the cytoplasm. One end of a transfer molecule contains a loop with three nucleotides positioned so that they will pair with three nucleotides in messenger RNA according to the base-pairing principle described earlier. The functions of transfer RNA are illustrated in Figure 1.7. The nucleotide triplets in the transfer RNA molecules are called *anticodons*.

While one end of a transfer RNA molecule contains the anticodon, the other end is designed to form an attachment with one of the approximately 20 amino acids that can be involved in protein synthesis. Each tRNA molecule will attach to only one specific amino acid. However, since 64 different molecules of tRNA are possible, some of the amino acids are associated with more than one transfer RNA. As the molecules of tRNA with their respective amino acids attached move into place along the molecule of messenger RNA, the amino acids are placed in position to be connected together to form a chain of protein. After the amino acids become attached to each other, the tRNA molecules are released both from the mRNA and from the amino acids. Thus the process whereby the sequence of nucleotides in a gene controls the sequence of amino acids in a protein is complete. The completed protein is now free to move on to carry out its prescribed function. In this example, the protein could be an enzyme that will be involved as a catalyst in a certain chemical reaction.

1.2.3 How Enzymes Control Phenotypes

Enzymes function as catalysts for chemical reactions. The functions of a cell involve hundreds and thousands of chemical reactions essentially all of which require specific enzymes to be carried out. Each enzyme is a specific protein for which a certain gene is necessary to control its synthesis. Figure 1.8 is a simplified illustration of how genes and their enzymes control chemical reactions, which in turn control phenotypes.

Phenylalanine is an amino acid that is a common component of many proteins. It is commonly found in the blood of most animals, including humans. One of the ways it is metabolized is by being converted to another amino acid, called *tyrosine*. This conversion requires the services of an en-

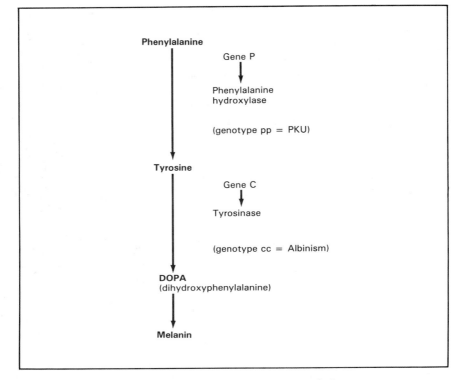

Figure 1.8   Illustration of how genes may control phenotypes.

zyme whose synthesis is under the control of a specific gene. If that gene is present, the enzyme will be synthesized and tyrosine will be produced. If the specific gene, which can be represented by the letter *P*, is missing, such as in the genotype *pp*, the appropriate enzyme will not be produced and phenylalanine will not be converted to tyrosine. In human babies with this condition, the improper metabolism of phenylalanine results in a disease called *phenylketonuria* (PKU), which is associated with serious mental retardation.

The amino acid tyrosine can be further metabolized to the pigment *melanin* under the control of an enzyme called tyrosinase. Melanin is the basic pigment which produces the various colors of hair, eyes, and skin. In the absence of the gene that controls the synthesis of tyrosinase, the enzyme is not produced and thus no melanin can be produced. This results in the absence of all color, a condition known as *albinism*.

Once melanin is produced, a number of other enzymes are necessary to produce the colors or combination of colors that characterize the many phenotypes of our domestic animals. The synthesis of each of these enzymes requires the services of a different gene, further illustrating how genes and enzymes play an important role in the control of phenotypes.

1.3 MITOSIS

To better understand how genes segregate in the formation of gametes, it is necessary to have a rather thorough understanding of how cells divide. Two types of cell division occur in mammals. One type involves a single division of a diploid cell to produce two identical diploid daughter cells. This kind of division is called *mitosis*. An understanding of mitosis will help in explaining the second kind of cell division, called *meiosis*, which involves a series of two divisions whereby four haploid daughter cells are produced from one diploid parent cell. Mitosis, the simpler of the two kinds of division, occurs among the body cells. Body cells, or somatic cells, include all cells except the sex cells, or gametes. Mitosis functions to increase the number of cells in growing animals and to maintain cell numbers in tissues as old cells wear out and die.

1.3.1 Interphase

In dividing cells, *interphase* represents the period of time from the end of one division to the beginning of the next. A number of events occur during this phase in preparation for the next division, but they cannot be seen through a microscope. The duration of interphase may be as short as a few hours or more than 100 hours, depending on the organism. The typical length of time in dividing cells is probably around 15 to 30 hours. Interphase is generally divided into three stages based on the activities that occur within the cell.

The first of the three stages of interphase is referred to as the *G1 stage*, or *first growth stage*. It has also been called the first gap between the end of the previous division and DNA synthesis of the next. It is a highly variable stage with regard to duration. It may be almost totally absent in rapidly dividing cells or may last for more than 100 hours in slowly dividing mature cells. The activities of this stage are related to preparing the cell for DNA synthesis and chromosome duplication. This probably involves some synthesis of RNA and enzymes involved in DNA synthesis. The other normal functions of the cell apparently continue to be carried out during this stage.

The next stage of interphase is called the *synthesis stage*, or *S stage*. It is during this time that each chromosome strand synthesizes a new strand parallel to itself and attached to a common centromere. The two identical strands are called *chromatids*. This is probably the most significant stage of the entire cycle of division. This is when the doubling process occurs that results in the identical genetic makeup of the two daughter cells of the mitotic process.

The third and final stage of interphase is called the *second growth stage*, the *G2 stage*, or the *post-DNA stage*. Generally speaking, the length of the G2 stage is probably shorter than the G1 stage. The relative duration of the three stages of interphase is quite variable depending on the organism, the type of cell, the maturity of the organism, and other factors. Figure 1.9 illustrates the three stages of interphase and the four phases of the mitotic process as it might occur in a typical cell.

1.3.2 Prophase

This phase of cell division includes most of the activities that take place to prepare the cell for actual division. It is usually defined as beginning when the chromosomes become shorter and thicker, such that they begin to be distinguishable when observed through a light microscope. Chromosome duplication will have already occurred during interphase. In advanced stages of prophase as observed through a microscope, the chromosomes have two strands attached by one centromere. A diagram of mitosis is shown in Figure 1.10.

The events included in prophase occur to bring about an orderly division of the replicated chromosomes so that each daughter cell will be assured of receiving one complete set of homologous pairs. As the chromosomes, called *chromatin* in the interphase cell, become distinguishable, several other events are also taking place. The centriole, positioned just outside the nuclear membrane in the interphase cell, divides and migrates toward opposite sides of the cell. As the centrioles migrate, the nuclear membrane seems to disappear. As the centrioles approach opposite sides of the cell, a spindle forms, seeming to connect one centriole to the other. The spindle appears as a series of fibers extending across the cytoplasm from

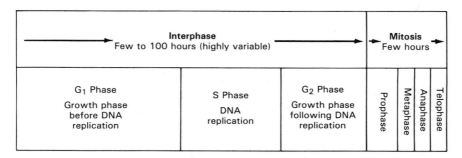

Figure 1.9 Interphase and mitotic cycle of a dividing cell.

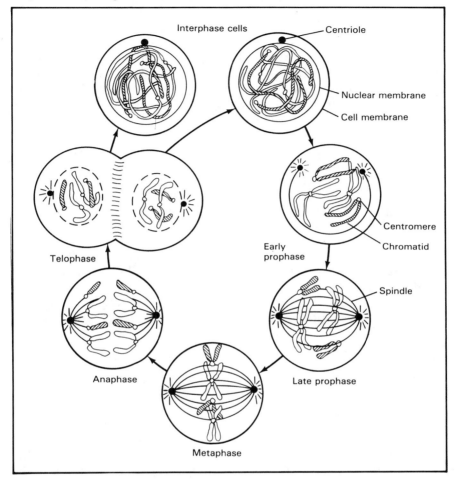

Figure 1.10 Cell undergoing mitosis. [From Charles E. Stufflebeam, *Principles of Animal Agriculture,* © 1983, p. 123. Reprinted by permission of Prentice-Hall, Inc., Englewood Cliffs, NJ.]

centriole to centriole. The fibers also appear to be connected to the centromeres of the doubled chromosomes. At this stage, the chromosomes are distributed at random throughout the cytoplasm with no obvious attraction or association between homologs.

1.3.3 Metaphase and Anaphase

The metaphase of mitosis is described as the phase wherein the chromosomes become arranged along the equatorial plane of the cell. The chromosomes are situated so that when the centromeres actually divide and begin to separate during anaphase, each new cell will be better assured of obtaining a complete complement of new chromosomes. During anaphase, the centromeres divide longitudinally so that the chromatids separate. The centromeres move toward the poles of the cell represented by the position of the centrioles, as though they were being pulled along by the spindle fibers. Metaphase and anaphase are both relatively short phases compared to the length of prophase.

1.3.4 Telophase

This is the final stage of the mitotic process. After the newly formed chromosomes have completed their migration, forming a cluster next to each centriole, the cell membrane constricts between them and eventually separates into two complete cells. As this constriction and division of the cell proceeds, the nuclear membrane forms around each cluster of chromosomes. The chromosomes again become long and slender and indistinguishable. The two new daughter cells return to the interphase. Both new cells are diploid and identical to each other and to the parent cell from which they were formed. The production of the two diploid cells from a single diploid cell is made possible by the fact that the chromosomes had duplicated prior to the actual division of the cell.

1.4 MEIOSIS

The second of the two major types of cell division occurs with specialized cells located in the testes and ovaries. The process begins with diploid cells called *gonial cells*, or *primordial germ cells*. In males these cells are called *spermatogonia* and are located inside a network of ducts called the *seminiferous tubules*. In females, the germ cells are called *ovigonia* (also *oogonia*) and are located just beneath the outer covering or germinal epithelium of the ovaries. Meiosis occurs as a part of the overall processes of gamete formation in animals: spermatogenesis in males, and ovigenesis (oogenesis) in females. The process of meiosis involves two divisions and results in the production of four haploid cells from one germ cell. Meiosis occurs only in connection with the production of male and female sex cells.

1.4.1 The First Meiotic Division

This division begins in much the same way as does mitosis. The events that take place in the interphase and prophase of mitosis also occur in preparation for the first division of meiosis. Again, probably the most significant event that occurs during this period of time from a genetic standpoint is the duplication of chromosomes. As in mitosis, this occurs during the interphase. During the prophase of the first division, the chromosomes become relatively short and thick, the centriole divides, the nuclear membrane disappears, and the spindle forms in much the same way as it did in the prophase of mitosis. Since there are two divisions in meiosis, this prophase is referred to as *prophase I.*

An additional event occurs during prophase I of meiosis that did not occur in mitotic prophase; this is the process of *synapsis.* During this phase, each pair of doubled homologous chromosomes are attracted one to the other and become arranged side by side, forming a four-stranded object called a *tetrad.* In a meiotic cell in prophase I in hogs, for example, there would be 19 tetrads since porcine body cells contain 19 pairs of homologous chromosomes. The process of synapsis is sometimes referred to as the process of tetrad formation. The first meiotic division is illustrated in Figure 1.11. It should be helpful to refer to these diagrams several times during the reading of this description.

During the next phase of the first division, the tetrads become arranged along the equatorial plane of the cell in a manner similar to that of the metaphase of mitosis. In meiosis, this is called *metaphase I.*

The events of *anaphase I* of meiosis are similar in some respects to those of the anaphase of mitosis, but there are also some differences. Recall that in mitosis, there were no tetrads; the homologous chromosomes were distributed randomly throughout the cell. During mitotic anaphase, the centromeres divided and the sister chromatids separated, each forming a new chromosome. In anaphase I of meiosis, this does not occur. Instead, the homologous pairs (tetrads) separate, or become "unsynapsed." The separated homologs move away from each other and move toward the centrioles at the poles of the cell. At this point, each homolog still contains two chromatids.

The events of *telophase I* are similar to those of mitotic telophase. A nuclear membrane forms around the cluster of chromosomes near each centriole. The cell membrane constricts between the two new nuclei and forms two new daughter cells. This completes the first meiotic division. The two daughter cells are haploid, having only one chromosome from each pair of homologs. Because of this reduction in chromosome number that occurs during the first division, it is often called the *reduction division.*

During prophase I and metaphase I, while the homologous chromosomes are in synapsis, a phenomenon called *crossing-over* takes place. A chromatid of one chromosome will lie across one of the chromatids of the homolog, forming an X-shaped connection between them. This crossover normally occurs at the same locus of the two nonsister chromatids, and is called a *chiasma*, for the Greek letter chi (χ). When the homologous chromosomes separate during anaphase I, the X-shaped connection must also

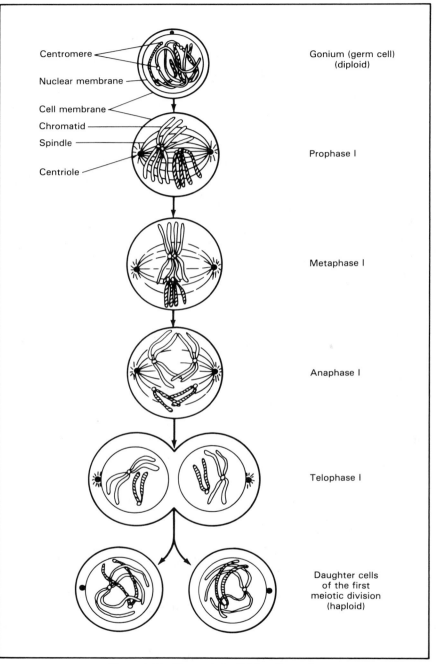

Figure 1.11   First meiotic division ($2n = 4$).

separate since it is between nonsister chromatids. When this separation occurs, the pair of homologs may very well exchange parts of their chromatids. Genes that were once on one chromosome may now be found on the homolog. This is a normal part of the meiotic process. Crossing-over plays an important role in the mixing of genes between homologs, and in the random distribution of genes in the gametes. Further discussion of crossing-over can be found in Chapter Nine.

1.4.2 The Second Meiotic Division

The second division, diagrammed in Figure 1.12, is also divided into phases to simplify its explanation. Prophase II involves several of the events that were described in mitotic prophase and in prophase I. The one centriole of each cell forms two, which then migrate to opposite sides of the cell; the nuclear membrane disappears; the chromosomes become short and thick; and a spindle forms. However, the chromosomes do not undergo another duplication, nor does synapsis occur again.

Metaphase II involves the orientation of the chromosomes on the equatorial plane of the cell in much the same way as it occurred in mitotic metaphase and metaphase I. Anaphase II is more like mitotic anaphase than it is like anaphase I. In anaphase II, the centromeres of each doubled chromosome divide and form two new centromeres, which then move toward opposite sides of the cell. The chromosomes, each of which have had two chromatids since the duplication process occurred in interphase, have now separated into two sets of single-stranded chromosomes. Telophase II completes the second division of meiosis, producing two haploid products. Because the second division produces haploid cells from haploid parent cells, it is often called *equational division*.

1.4.3 Gametogenesis and Fertilization

In males, as a spermatogonium prepares for the first division, all of the events described in interphase and prophase I take place. As the cell undergoes these preparations, it also increases in size. At this point the meiotic cell is called a *primary spermatocyte*. The division of a primary spermatocyte produces two haploid cells called *secondary spermatocytes*. The products of the second meiotic division are called *spermatids*. The changes that occur to produce a sperm cell from a spermatid are not genetic changes. During this process, most of the cytoplasm of the spermatid is lost and a tail develops. Spermatogenesis is illustrated in Figure 1.13.

In females, the process of ovigenesis takes place inside specialized structures on the ovaries called *follicles*. As ovigenesis progresses in mam-

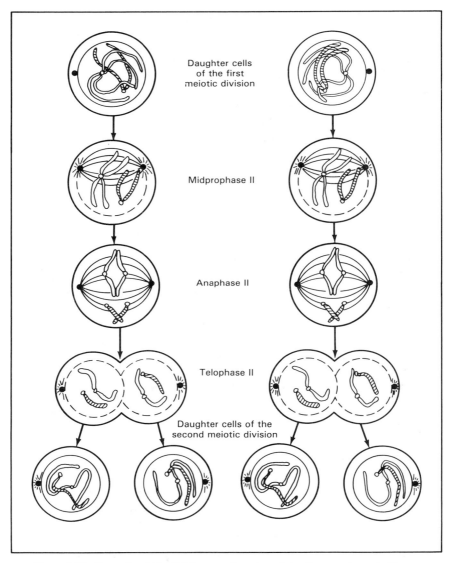

Daughter cells
of the first
meiotic division

Midprophase II

Anaphase II

Telophase II

Daughter cells of the
second meiotic division

Figure 1.12   Second meiotic division; continuation of the diagram shown in Figure
1.11.

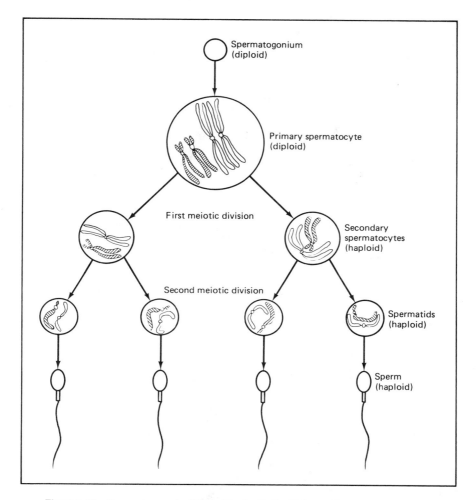

Figure 1.13   Spermatogenesis. [From Charles E. Stufflebeam, *Principles of Animal Agriculture,* © 1983, p. 125. Reprinted by permission of Prentice-Hall, Inc., Englewood Cliffs, NJ.]

mals, the follicles enlarge in size to about $\frac{1}{2}$ inch in diameter or larger, depending on the species involved. As the events of prophase I occur, the ovigonium increases in size from that of a typical body cell to about 150 to 200 micrometers in diameter. During the process, it develops a relatively thick covering, called the *zona pellucida*, to protect and hold the cell together. There is a second, relatively thin membrane located just beneath the zona pellucida called the *vitelline membrane* (see Figure 1.14). During this

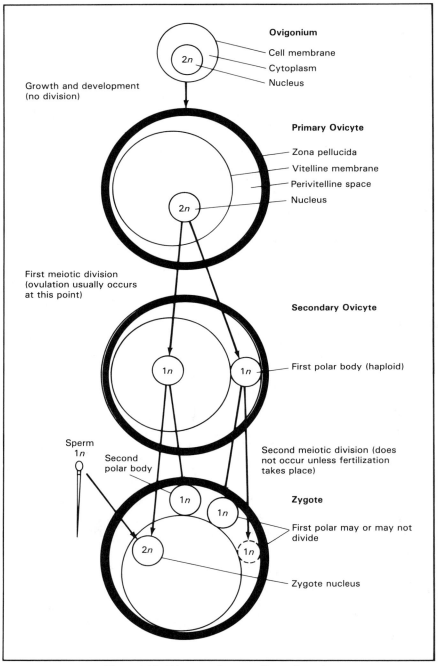

Figure 1.14 Diagram of mammalian ovigenesis.

period of time and up until the first division, the cell is called a *primary ovicyte.*

As anaphase I and telophase I occur within the primary ovicyte, two haploid nuclei are produced. One of these nuclei is retained within the cytoplasm of the cells; the other is expelled or extruded through the vitelline membrane into the space between the vitelline membrane and the zona pellucida. This extruded nucleus is called a *polar body* and serves no known genetic function. In fact, it usually disintegrates in a relatively short period. The remaining nucleus, along with the rest of the cell, is now called a *secondary ovicyte.*

In most species of mammals, as illustrated in Figure 1.14, the second meiotic division generally does not occur unless a sperm cell finds its way into the cell. The presence of the sperm cell sets up a reaction that stimulates the second meiotic division to take place. Again, the nucleus divides, producing two nuclei, one of which is extruded into the space between the vitelline membrane and zona pellucida. The expelled nucleus is called the second polar body and also has no further practical function. The nucleus that remains inside the cell ultimately unites with the nucleus of the sperm cell, producing the zygote. The processes of ovigenesis and fertilization are now both complete.

In the event that fertilization of the secondary ovicyte had not occurred, meiosis would not have progressed any further, and the cell would have died and become reabsorbed by the body.

STUDY QUESTIONS AND EXERCISES

1.1. Define each term.

allele	gene	mutation
anaphase	genetics	nucleotide
autosomes	genotype	oogenesis
centriole	haploid	oogonium
centromere	heredity	ovigenesis
chromatid	heterozygous	ovigonium
chromatin	homologous	ovum
chromosome	homozygous	phenotype
codon	hydrogen bonding	prophase
deoxyribonucleic acid	interphase	ribonucleic acid
deoxyribonucleotide	karyotype	ribonucleotide
deoxyribotide	locus	ribosome
diploid	meiosis	ribotide
DNA	metaphase	rRNA
enzyme	mitosis	sperm
fertilization	mRNA	spermatid

spermatogenesis	spermiogenesis	tetrad
spermatogonium	synapsis	tRNA
spermatozoon	telophase	variation

1.2. What are two major causes of variation in farm animals?

1.3. When and how did the science of genetics begin?

1.4. Describe the four categories of chromosomes with respect to the position of their centromeres.

1.5. What possible consequence can occur to an organism that has more or less than the usual diploid number of chromosomes in its body cells?

1.6. Rank the several species of domestic animals according to the number of chromosomes common in their body cells.

1.7. Name the four kinds of nucleotides found in DNA.

1.8. How does the basic structure of a ribonucleotide differ from that of a deoxyribonucleotide?

1.9. Briefly describe the base-pairing theory with respect to the structure of DNA.

1.10. Explain how a chromosome duplicates itself.

1.11. What determines the sequence of amino acids in any particular protein?

1.12. Compare the structure of RNA to that of DNA.

1.13. What is the function of messenger RNA?

1.14. How many kinds of transfer RNA can be found in a cell, and what determines that number?

1.15. Why is it necessary for the chromosomes in a cell to duplicate before mitosis?

1.16. What is the significance of mitosis; in other words, why does it occur?

1.17. Compare mitotic prophase to prophase I of meiosis.

1.18. What is the significance of crossing-over with respect to meiosis?

1.19. How does anaphase I of meiosis differ from mitotic anaphase?

1.20. What is the significance of the terms *reduction division* and *equational division*?

1.21. What is sexual reproduction?

PROBLEMS

1.1. Identify the specific phase of mitosis or meiosis for each of the following diagrams ($2n = 6$).

(a)   **(b)**

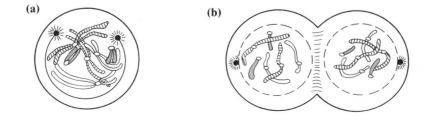

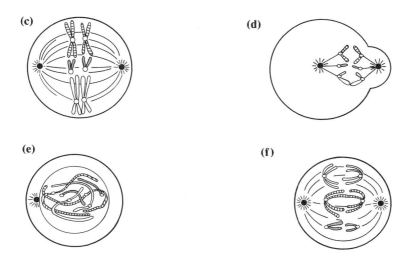

1.2. Assume that a hypothetical organism has three pairs of homologous chromosomes in its body cells: one pair of metacentric, one pair of submetacentric, and one pair of telocentric chromosomes. Prepare sketches or diagrams of each of the following cells.

(a) a cell in mitotic prophase

(b) a cell in mitotic anaphase

(c) primary ovicyte in prophase I

(d) secondary spermatocyte in anaphase II

(e) spermatid

(f) zygote

1.3. Suppose that the three pairs of chromsomes in Problem 1.2 are labeled *A/a, B/b,* and *C/c.* How many different kinds of gametes can be produced from this individual? List them.

1.4. Of the 38 chromosomes of hogs, state how many would be found in each of the following.

(a) primary spermatocyte

(b) spermatid

(c) ovigonium

(d) polar body

(e) secondary spermatocyte

(f) spermatogonium

(g) sperm cell

(h) zygote

1.5. Name the stage in a mitotic somatic cell where each of the following events occurs. (a) Centromeres separate and chromosomes migrate to the centrioles. (b) Centrioles divide and migrate to opposite sides of the cell. (c) Chromosomes become oriented on the equatorial plate. (d) Nuclear membrane disappears. (e) Chromosomes become shorter and thicker. (f) Chromosomes replicate.

REFERENCES

BURNS, G. W. 1980. *The Science of Genetics,* 4th ed. Macmillan Publishing Company, Inc., New York.

ELDRIDGE, F. E. 1985. *Cytogenetics in Livestock.* AVI Publishing Company, Westport, Conn.

LASLEY, J. F. 1987. *Genetics of Livestock Improvement.* Prentice-Hall, Inc., Englewood Cliffs, N.J.

MENDEL, G. 1865. *Experiments in Plant Hybridization.* Reprinted in C. Stern and E. Sherwood, eds. 1966. *The Origin of Genetics—A Mendel Source Book.* W.H. Freeman and Company, Publishers, San Francisco.

WATSON, J. D., and F. H. C. CRICK. 1953. *Genetical Implications of the Structure of Deoxyribonucleic Acid.* Reprinted in J. A. Peters, ed. 1959. *Classic Papers in Genetics.* Prentice-Hall, Inc., Englewood Cliffs, N.J.

PART 2
QUALITATIVE INHERITANCE

TWO

One Pair
of Genes

In Chapter One we learned about the nature and function of genes and chromosomes and how they are able to reproduce themselves. We also saw how chromosomes are distributed to the gametes in such a way that only one chromosome from each pair of homologs occurs in the gametes. In this chapter we deal with genes and alleles one pair at a time and how they segregate and recombine to produce certain results. The principles that apply to one pair of genes will usually apply to many pairs of genes.

2.1 THE DISTRIBUTION OF GENES FROM GENERATION TO GENERATION

2.1.1 Segregation and Recombination of Genes

A natural law exists that governs the distribution of genes from one generation to the next. This law employs the rules of probability. It may be referred to by any one of several names. It may be called the *law of random assortment of genes*, the *law of segregation and recombination*, or the *first law of genetics*. Some refer to it as *Mendel's first law* in honor of the Austrian monk who discovered it and published his findings in 1865. The entire subject of segregation and recombination of genes and their expression in phenotypes is sometimes referred to as *Mendelian genetics*. Regardless of the terminology used, the law involves two basic parts: segregation of genes in the gametes and recombination of genes in the zygote.

Genes segregate in the gametes during the process of meiosis in spermatogenesis and ovigenesis. The two members of each pair of homologous chromosomes are distributed to a separate gamete. Except in rare cases of abnormal meiosis, two homologous chromosomes are never found in the same gamete. Therefore, the two genes found at any two loci on a pair of homologous chromosomes are never found in the same gamete. The genes will segregate in the gametes in all possible combinations due to chance. Individuals of the genotype AA will produce only one kind of gamete, the kind containing one A gene. In similar fashion, the aa genotype will produce only a-bearing gametes. From this bit of information, the conclusion can be made that homozygous individuals (homozygotes) produce only one kind of gamete. On the other hand, heterozygotes can produce two kinds of gametes for every pair of heterozygous genes they possess. If we are considering the genes at only one locus, the individual of the genotype Aa will produce only two kinds of gametes, those containing an A and those containing the allele a.

During the mating process, when the gametes from the male and female unite, the diploid number of chromosomes is once again restored in the zygote. Genes are again recombined into pairs. This recombination process results in one or more possible kinds of genotypes. The genotype resulting from the process of recombination depends on the genes carried in the gametes, which in turn depend on the genotypes of the parents.

2.1.2 The Six Basic Crosses

When dealing with any trait in a population that is controlled by two alleles, A and a, as we have seen, three different genotypes are possible: AA, Aa, and aa. The three genotypes can be united in crosses in six possible ways, as shown in Table 2.1.

TABLE 2.1   Six Basic Crosses Involving One Pair of Genes

Parents	Progeny
1. $AA \times AA$	All AA
2. $AA \times aa$	All Aa
3. $aa \times aa$	All aa
4. $Aa \times AA$	$\frac{1}{2} Aa, \frac{1}{2} AA$
5. $Aa \times aa$	$\frac{1}{2} Aa, \frac{1}{2} aa$
6. $Aa \times Aa$	$\frac{1}{4} AA, \frac{1}{2} Aa, \frac{1}{4} aa$

The six crosses can be categorized into three groups, based on the numbers of kinds of progeny produced. The first three crosses are similar in at least two ways: all individuals being mated are homozygous, and only one kind of progeny is produced from each cross. Since homozygotes can produce only one kind of gamete, those gametes can unite at fertilization in only one way. This is shown in the following three illustrations:

Cross 1
genotypes of the parents

gametes of the parents
genotype of the progeny

Cross 2
genotypes of the parents

gametes of the parents
genotype of the progeny

Cross 3
genotypes of the parents

gametes of the parents
genotype of the progeny

Based on the results of these three crosses, it can be concluded that matings between any two homozygous individuals will produce *only one kind of genotype* among their progeny.

Two of the crosses, 4 and 5, are similar in that one individual is heterozygous and one is homozygous, and that two kinds of progeny are produced from each cross. In addition, the genotypes of the two kinds of progeny are identical to those of the parents, as illustrated in cross 4:

genotypes of the parents
gametes of the parents
genotypes of the progeny

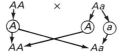

Based on the results of crosses 4 and 5, it can be stated that the mating of a heterozygote with a homozygote will produce progeny one half of which will be heterozygous and one half of which will be homozygous. The genotypes and genotypic proportions among the progeny will be the same as those of the parents.

Cross 6 is in a category all its own. It involves a mating of two heterozygous individuals. As has been the case with each of the other five crosses, this cross also illustrates how the genes segregate in the gametes and recombine in the progeny in all possible combinations due to chance.

| genotypes of the parents | 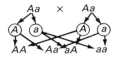 |
| genotypes of the progeny | |

The progeny of this cross are distributed in the ratio $\frac{1}{4}AA : \frac{1}{2}Aa : \frac{1}{4}aa$. Stating it in another way, the genotypic ratio of a cross between two heterozygotes is $1 : 2 : 1$. This ratio represents the numerators of the fractions $\frac{1}{4}$, $\frac{2}{4}$, and $\frac{1}{4}$. This particular cross is special in that all possible kinds of genotypes that can exist in a system with only two alleles are present among the progeny. For this reason, when studying the ways in which genes express themselves in phenotypes, an attempt should be made to obtain this particular mating.

An individual heterozygous for one pair of genes (Aa) is called a *monohybrid*. The prefix *mono-* refers to one pair of genes; the suffix *-hybrid* refers to the fact that those genes are heterozygous. In genetics the word *hybrid* generally means heterozygous or heterozygote. A cross of two individuals that are heterozygous for one pair of genes is called a *monohybrid cross*.

Essentially all of the crosses dealing with one pair of alleles will involve one or more of the six basic crosses. The principles dealing with these crosses will be used later in the solution of crosses involving two or more pairs of genes. Because of the importance attached to the six basic crosses, especially the monohybrid cross, it is recommended that the crosses and their genotypic results be committed to memory.

2.2 DOMINANT AND RECESSIVE GENES

In dealing with phenotypes and the genotypes that control them, we must keep in mind that most of the time, the effects of genes must be considered two at a time. If the two genes in any genotype are the same, as in AA and aa, the effects of those genes will be the same upon the phenotype. However, when the heterozygote is involved, we must consider the effects of both of those alleles. Studying the effect of the heterozygous genotype on the phenotype lets us decide how the two alleles interact. If the phenotype of the heterozygote (or hybrid) is identical to one of the homozygotes, we say that the gene that is present in that particular homozygote is completely *dominant* over its allele. The allele in the heterozygote that is not expressed is called *recessive*. In most situations when the term *dominant* or *dominance* is used, it will mean complete dominance unless otherwise specified.

2.2.1 Crossing the Purebreds

Poultry of the White Leghorn breed have a single comb, while rose comb is normally present in birds of the White Wyandotte breed. The rose comb is short and has a folded appearance that in some way resembles a rosebud (see Figure 2.1). The rose-combed trait is the result of the presence of a dominant gene which we will symbolize with the letter R. Since it is dominant, the rose-comb phenotype can result from either the homozygous (RR) or heterozygous (Rr) genotypes. The single comb will result only from the homozygous state of the recessive allele (rr). Purebred White Wyandotte birds are more often homozygous for the rose comb than they are heterozygous. The problem of the heterozygote will be dealt with later. Assuming homozygosity for the dominant phenotype, crosses between the two breeds of poultry will produce rose-combed progeny which are heterozygous ($RR \times rr =$ all Rr). This cross is the same as cross 2 of the six basic crosses shown in Table 2.1. The cross between these two purebreds is called a *parental* or P_1 *cross*. The heterozygous progeny are referred to as the first *filial* or F_1 *generation*.

2.2.2 The Monohybrid Cross

A cross between two heterozygous rose-combed birds constitutes the monohybrid cross. To be sure that the rose-combed birds are indeed heterozygous, birds should be chosen that have one homozygous recessive parent, as demonstrated in the previous paragraph. The genotypic ratio of the mo-

Figure 2.1   Rose-comb and single-comb types in chickens. Left; single comb; right; rose comb.

nohybrid cross for these alleles is $1RR : 2Rr : 1rr$. The two genotypes having the dominant gene R will express the rose-comb phenotype. Therefore, the phenotypic ratio of the monohybrid cross when dominace is involved is three dominant phenotypes to one recessive. The progeny produced from mating two F_1 individuals together are called the F_2 *generation*. In this case, as shown in Figure 2.2, three rose-combed birds are produced from every single-combed bird. Slight deviations from this ratio can be expected in ac-

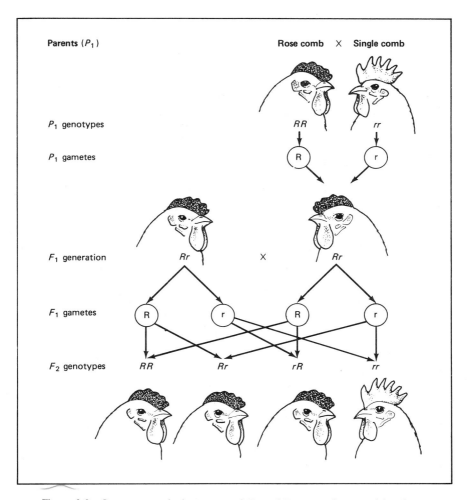

Figure 2.2 Genotypes and phenotypes of F_1 and F_2 generations resulting from crosses between homozygous rose-combed and single-combed chickens. [From Charles E. Stufflebeam, *Principles of Animal Agriculture,* © 1983, p. 133. Reprinted by permission of Prentice-Hall, Inc., Englewood Cliffs, NJ.]

tual crosses. Assume that from several monohybrid matings 100 chicks were produced. According to the theoretical ratio, 75 rose-combed chicks and 25 single-combed ones would be expected. However, deviations of up to four or five in each group would not be unusual in a population of this size. A statistical test called the *chi-square test* can be applied to determine how significant certain deviations from the expected results may be. This test is explained in Chapter Four.

2.2.3 The Testcross

As we have already seen, when dominance is involved between alleles, a problem exists as to our ability to distinguish between the dominant homozygote (*RR*) and the heterozygote (*Rr*). In situations such as this, the notation R_ can be used to represent both of the dominant genotypes. The dash indicates that either the dominant or the recessive allele could be present as the second gene, but we do not know which it is. A breeding test called the *testcross* can be performed to help determine which of the two genotypes is present in the dominant phenotype. Suppose that a Wyandotte breeder is interested in obtaining the services of a new cock (rooster) to breed to his hens and does not want to risk the chance of producing some heterozygous chicks. Mating this cock to several single-combed hens would provide the breeder with information to determine the genotype of the cock before breeding him to his hens. As shown in the following illustration, if the cock were heterozygous, mating him to single-combed hens should produce progeny one-half of which would be single-combed and one-half rose-combed:

Parents: Rose-combed male (*Rr*) Single-combed female (*rr*)
Progeny: $\frac{1}{2}$ rose-combed chicks (*Rr*)
           $\frac{1}{2}$ single-combed chicks (*rr*)

The appearance of any single-combed chicks would be genetic proof of heterozygosity in the sire. On the other hand, if the cock were homozygous, no single-combed chicks would be expected. Since the possibility exists that he could be heterozygous and still sire several rose-combed chicks without the appearance of any single-combed ones, 5 or 10 or more progeny should be observed before deciding that he is quite probably homozygous. The probability that any one of his progeny would be rose-combed if he were heterozygous is $\frac{1}{2}$. The probability that he might produce 5 progeny with rose combs is $\frac{1}{32}$, or 1 chance in 32. The production of 10 progeny with rose combs under these conditions would be expected only once in 1024 times $(\frac{1}{2})^{10}$. The logical conclusion to be drawn if this cock should sire as many as 10 rose-combed progeny, and none with single combs, is that he is almost surely homozygous. More will be presented about testing for the presence of recessive genes in Chapter Four.

Obtaining sufficient testcross progeny from either male or female birds, or from male or female litter-bearing mammals, can normally be done with one mating since between 5 and 10 progeny are usually enough to determine the genotype of an individual expressing a dominant trait with a reasonable degree of assurance. With cattle, horses, and sheep, even though several matings would be needed to obtain the necessary 5 to 10 progeny, males could be properly tested in one breeding season. In monotocous females (those with normally single births), however, it would require several years to obtain the necessary numbers of progeny for a proper test unless embryos were transferred. For this reason, it is usually not practical to test cows, mares, or ewes for the presence of recessive genes.

2.2.4 Examples of Dominant and Recessive Traits

The example of comb types in chickens is an excellent illustration of the principles of dominance and recessiveness between alleles. Many other examples in domestic animals can be given. Some of these example will be used in succeeding chapters and in the problems at the ends of chapters. Table 2.2 is only a partial listing of dominant and recessive traits in some selected domestic animals.

2.3 MODIFICATIONS OF MONOHYBRID RATIOS

2.3.1 Intermediate Degrees of Allelic Interaction

Figure 2.3 illustrates graphically several degrees of allelic interaction. It shows that when two alleles are present together in the same genotype, they do not always interact with one allele exerting complete dominance over the other. The roan color pattern in the Shorthorn breed of cattle illustrates this very well. Cattle of this breed may express any one of three colors: red, roan, or white. Red results from the homozygous state of one allele (RR), white is produced by the opposite allele when in the homozygous state (rr). The roan pattern is the result of the heterozygous condition (Rr). There is no dominance between the two alleles. The roan pattern is made up of both red hairs and white hairs growing among one another (Figure 2.4). The amount of roan varies from a few red hairs on an almost all-white animal to a few white hairs growing on an almost all-red individual. For this reason there is the possibility that some roan cattle may be recorded as white, while others may be recorded as red.

**TABLE 2.2   Some Examples of Dominant and Recessive Traits
in Selected Domestic Animals**

Species	Dominant Trait	Recessive Trait
Cattle	Black hair coat	Red hair coat
	Polled (no horns)	Horns
	White face	Solid color
	Solid color	Irregular white spotting
	Red	Yellow (diluted red)
	Cloven hooves	Mulefeet
Chickens	Rose comb	Single comb
	White skin	Yellow skin
	Dominant white	Color
	Colored feathers	Recessive white
	Pea comb	Single comb
	Feathered shanks	Clean shanks
	Crested head	No crest
	Black feathers	Red feathers
Horses	Black hair coat	Chestnut or sorrel
	Bay	Nonbay (black)
	Chestnut mane and tail	Flaxen mane and tail
	Smooth hair	Curly hair
Sheep	Hairy fleece	Wooly fleece
	White wool	Black wool
	Brown eyes	Blue eyes
Swine	Black hair (Hampshires)	Red hair
	White belt	No belt
	Erect ears	Drooping ears
	Mule foot	Cloven hoof
Dogs	Wire hair	Smooth hair
	Black hair	Liver color
	Red hair	Yellow hair
	Solid color	White spotting
Cats	Short hair	Long hair
	Black hair	Brown hair
	Agouti (wild color)	Nonagouti
	Color	White

Geneticists are not in complete agreement as to the terminology that should be used to describe this type of allelic interaction. The terms used most often are incomplete dominance or codominance. Other terms that have been used are lack of dominance, partial dominance, and blending inheritance. Whether the two alleles exert equal dominance upon one another, whether there is complete absence of dominant effects, or whether there are varying degrees of partial dominance between them, one thing is evident—a separate phenotype exists for each genotype, including the heterozygote.

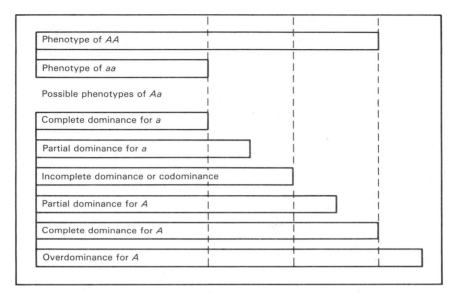

Figure 2.3   Graphic illustration of the various degrees of allelic interaction.

Figure 2.4   The roan pattern in a shorthorn cow. (Owner: John Sparkman, Republic, MO.)

In the case of the roan pattern in cattle, the white hairs are apparently no different than the white hairs of all-white cattle. Similarly, the red hairs of roan cattle are apparently the same as the red hairs of red cattle. This would indicate that each allele has the same effect in the heterozygote as in its respective homozygote. The term *codominance* would seem to be an appropriate term to describe this kind of inheritance. The fact that the ratio of red to white hair varies so much could well be due to the influence of other genes. Table 2.3 shows the expected results of the six possible kinds of matings among Shorthorn cattle.

The inheritance of the palomino color in horses is controlled by a pair of alleles that lack dominance between them. The chestnut or sorrel color is due to the homozygous state of one of the alleles. The allele of the chestnut gene is called a *dilution gene*. The presence of one dilution gene results in a copper or golden color called palomino. The homozygote composed of two dilution genes expresses an off-white or light cream color referred to as cremello. A more detailed description of the inheritance of coat colors in horses is given in Chapter Eight. Several examples of codominant inheritance are listed in Table 2.4.

Use of the term *partial dominance* would suggest that phenotypic value of the heterozygote, if it can be measured quantitatively, is not exactly at the midpoint between the values of the phenotypes of the two homozygotes (see Figure 2.3). To some, the term *incomplete dominance* may suggest that one allele is more dominant than the other. Naturally, if one allele is partially or incompletely dominant, the other allele would be partially or incompletely recessive.

In the application of these intermediate kinds of inheritance to the breeding of domestic animals, the terminology used may not be as important as knowing that the expected phenotypic ratios of monohybrid crosses should be 1:2:1, as opposed to the 3:1 ratios expected from the monohybrid crosses with complete dominance.

TABLE 2.3   **Phenotypic Results of Codominant Inheritance in Shorthorn Cattle**

Phenotype of Parents	Phenotypes of Progeny
Red (RR) × red (RR)	All red (RR)
White (rr) × white (rr)	All white (rr)
White (rr) × red (RR)	All roan (Rr)
Roan (Rr) × red (RR)	$\frac{1}{2}$ roan (Rr): $\frac{1}{2}$ red (RR)
Roan (Rr) × white (rr)	$\frac{1}{2}$ roan (Rr): $\frac{1}{2}$ white (rr)
Roan (Rr) × roan (Rr)	$\frac{1}{4}$ red (RR): $\frac{1}{2}$ roan (Rr): $\frac{1}{4}$ white (rr)

TABLE 2.4   Examples of Traits in Domestic Animals Controlled by Alleles Showing No Dominance between Them

Species	Homozygous Traits		Heterozygous Traits
Cattle	Red	White	Roan
Chickens	Black feathers	White	Blue
	Normal feathers	Extremely frizzled	Mild frizzling
Horses	Chestnut	White (cremello)	Palomino
	Bay	White (cremello)	Buckskin
Humans	M blood group	N blood group	MN blood group
Sheep	Normal wool (crimped)	Halo hairs	Hairy wool
Swine	Black (Berkshires)	Red	Red and black spots
Guinea pig	Yellow	White	Cream
Goats	Long ears	Short ears	Medium-length ears

2.3.2 Effects of Lethal and Detrimental Genes

A *lethal gene* is one whose effect on an animal is so drastic as to cause its death. The lethal effect may occur any time after fertilization; however, in most references to lethal genes, death usually occurs during embryonic development or during a period of time up to shortly after parturition or hatching. Technically, a gene can be called lethal if it causes the premature death of the animal that carries it. A gene that causes death sometime after birth or later in life may be called *sublethal*, *semilethal*, or *latent lethal*.

A lethal gene that is completely recessive must be in the homozygous state to be expressed, so unless the gene were an extremely latent lethal, the recessive homozygote would not live long enough to reproduce. The only way that a recessive homozygote could be produced would be by the mating of two heterozygotes (often referred to as *carriers*). The monohybrid cross would produce genotypes in the ratio of 1:2:1, as would any other mono-hybrid cross. However, since the recessive homozygote would die before birth or hatching, or shortly thereafter, the only phenotype that would exist would be the dominant normal phenotype (either *AA* or *Aa*).

Lethal genes result from mutations of normal genes. If a recessive gene mutates to a completely dominant gene, the new mutant would be expected to kill the individual carrying the gene before that individual had a chance to reproduce. The only way a dominant lethal gene could be transmitted would be if it were a latent or sublethal gene. The disease in humans called Huntington's chorea, which results in a progressive degeneration of the nervous system, is caused by a dominant gene that usually results in death during the fourth or fifth decade of life. It is a classic example of a dominant latent lethal gene.

Some lethal genes are neither completely dominant nor recessive, but show codominant or blending effects with their alleles. Several examples of this kind of lethality can be observed among domestic animals. One such example is the creeper trait in poultry. Creeper birds have shortened and deformed legs that cause them to appear to "creep." When normal birds are mated together, only normal progeny are produced. This would indicate that the normal phenotype is produced by a homozygous genotype, but it says nothing about the relationship between the alleles.

When creeper birds are mated to normal birds, one-half of the progeny are normal and one-half are creepers. This 1:1 ratio of progeny indicates that one of the parents is homozygous and one is heterozygous. Since it has been shown that normal birds are homozygous, the heterozygotes must then be the creepers.

To be sure that the creeper is indeed heterozygous, creepers should be mated among themselves. This would represent a monohybrid cross. If sufficient numbers of progeny were produced, the expected genotypic ratio would be the usual 1:2:1. If the creeper trait was due to a completely dominant gene, three-fourths of the progeny should be creepers and one-fourth should be normal. The actual results of such crosses are consistently two-thirds creepers and one-third normal. In addition, it is observed that hatchability is usually reduced by about 25 percent, indicating that a lethal gene may be involved.

The creeper condition is thus caused by the interaction between a lethal gene and the allele that controls the normal phenotype. The lethal gene is neither completely dominant nor completely recessive. The question then arises as to the appropriate terminology to be used to describe this kind of allelic interaction. Depending on the preferences of the reader, terms such as *incomplete dominance, partial dominance, codominance, blending inheritance*, or *intermediate inheritance* may be used. The allelic relationship is similar to that which controls the roan pattern in Shorthorn cattle and the palomino color in horses. The use of the terms *dominant* and *recessive* without any qualifying terms usually means *complete* dominance and *complete* recessiveness. It therefore seems clear that to refer to the lethal gene in this case as either a dominant lethal or a recessive lethal is incorrect. Several other examples of traits controlled by lethal genes lacking complete dominance or recessiveness are listed in Table 2.5.

On some occasions the embryo containing the lethal gene in the homozygous state is able to survive until after birth, but dies shortly thereafter. For example, when platinum foxes are mated together, a pure white pup will occasionally be produced in addition to the expected platinum and sil-

TABLE 2.5   Some Traits Controlled by Lethal Genes Lacking Complete Dominance or Recessiveness

Species	Heterozygote	Homozygote
Chickens	Creeper	Normal
Sheep	Gray fleece	Black
Mouse	Yellow	Nonyellow
Fox	Platinum	Silver
Cattle	Comprest dwarf	Normal
Horse	Roan	Nonroan
Dog	Hairless	Normal
Rabbit	Pelger (abnormal WBC)	Normal

ver pups, but it usually dies within a few hours. The pure white pup must certainly be the lethal homozygote. When crosses between comprest dwarfs (referred to in Table 2.5) are successful, on certain occasions a highly abnormal dwarf called a bulldog is produced, but it usually lives only a few hours. The occurrence of the bulldog dwarf can be explained in a manner similar to that of the pure white fox pup.

STUDY QUESTIONS AND EXERCISES

2.1. Define each term.

blending inheritance	incomplete dominance	progeny
codominance	latent lethal gene	purebred
dominance	lethal gene	recessiveness
F_1	monohybrid	recombination
F_2	monohybrid cross	segregation
filial	P_1	sublethal
genotype	parental cross	testcross
heterozygote	partial dominance	
homozygote	phenotype	

2.2. What significance is attached to use of the term *Mendelian genetics*?

2.3. What is the first law of genetics?

2.4. With which biological event is gene segregation associated?

2.5. With which biological event is gene recombination associated?

2.6. Explain the rule of thumb that applies to the results of crosses between two homozygous individuals.

2.7. Explain the rule of thumb that applies to the results of crosses between a homozygote and a heterozygote.

2.8. What is the significance of the monohybrid cross?

2.9. How many kinds of gametes can be produced by a homozygote?

2.10. How many kinds of gametes can be produced by an individual heterozygous for one pair of genes?

2.11. Suggest at least two ways to be sure that a heterozygote expressing a dominant trait is indeed heterozygous.

2.12. What can be done to determine whether an individual that expresses a dominant trait is heterozygous or homozygous?

2.13. Explain the concept of allelic interaction.

2.14. What distinction, if any, can be made among the terms *codominance, partial dominance*, and *blending inheritance*.

2.15. Is it possible for a dominant lethal gene to become established in a population? Explain.

2.16. Is it possible for a recessive lethal gene to become established in a population? Explain.

2.17. Explain why the gene causing the creeper condition in poultry cannot be described as being dominant.

PROBLEMS

2.1. A purebred Irish Terrier with wire hair was bred to two Boston Terrier bitches with smooth hair and produced two litters of wire-haired pups. The dogs from these two F_1 litters were eventually mated together, producing 31 wire-haired and 11 smooth-haired F_2 progeny.

(a) Which of the two traits is controlled by a recessive gene?

(b) How many of the 31 wire-haired F_2 pups would be expected to be homozygous?

(c) How many of the 11 smooth-haired F_2 pups would be expected to be homozygous?

(d) How could you determine which of the wire-haired F_2 pups were heterozygous and which were homozygous?

(e) If smooth-haired dogs were mated together, would you ever expect any of their pups to have wire hair? Why?

(f) If wire-haired dogs were mated together, would you ever expect any of their pups to have smooth hair? Why?

2.2. Suppose that a group of white chickens mated among themselves always produced progeny with white feathers. What can you conclude about the genetic nature of this trait?

2.3. The albino trait in humans is caused by the homozygous state of a recessive gene (*aa*); normal pigmentation is due to the dominant allele (*A*).

(a) An albino man and his pigmented wife have a normally pigmented son; what are the genotypes of these three people?

    (b) Another albino man and his wife, who is normally pigmented, have six normally pigmented children; what is the probable genotype of the wife?

    (c) A husband and wife, who are both normally pigmented, have an albino child. If they should eventually have other children, what is the chance that some of them would be albinos?

2.4. Yellow guinea pigs mated together always produce yellow offspring; yellow animals mated to cream-colored guinea pigs produce these two colors in the progeny in a 1:1 ratio.

    (a) What kind of interaction exists between the alleles affecting these colors?

    (b) Which color is produced by the heterozygote?

2.5. In addition to yellow and cream, white is also controlled by the alleles in this series (number 4); what phenotypes and ratios would be expected among the progeny of these crosses? **(a)** yellow × white; **(b)** cream × cream; **(c)** white × white; **(d)** white × cream.

2.6. A pair of codominant alleles in humans, M and N, produce the three blood groups M (genotype MM), MN (heterozygous), and N (genotype NN).

    (a) Which blood groups would you expect to find among the children where the father is group M and the mother is group N?

    (b) Is it possible for a man with group M blood to be the father of a child who has group N blood? Explain.

2.7. The palomino color in horses is produced by the interaction between codominant alleles that in the homozygous state produce chestnut (CC) and white (cc). Assuming that a sufficient number of foals would be produced to obtain all possible combinations, what phenotypes (including ratios) would you expect from each of these crosses? **(a)** chestnut × chestnut; **(b)** white × white; **(c)** white × chestnut; **(d)** chestnut × palomino; **(e)** white × palomino; **(f)** palomino × palomino.

2.8. Is it possible to develop a breed of horses that would consistently transmit the palomino color? Explain.

2.9. Is it possible to devise a mating system that would always be expected to produce palomino foals? Explain.

2.10. The hairless trait as it occurs in Mexican Hairless dogs is produced by the heterozygous genotype (Hh). The normal condition is homozygous (hh). The H allele in the homozygous state results in the death of the embryo.

    (a) What phenotypic results would be expected from mating hairless dogs to normal dogs?

    (b) What would the phenotypic results be from mating hairless dogs among themselves?

2.11. The creeper trait in chickens results from the interaction of two alleles in the heterozygous state (Cc). The homozygote (cc) produces normal birds, while the other homozygote (CC) results in dead embryos and eggs that do not hatch. If hatchability is 80 percent when mating normals among themselves, what would the expected percent hatchability be from the eggs produced from creepers mated to creepers?

REFERENCES

BURNS, G. W. 1980. *The Science of Genetics*, 4th ed. Macmillan Publishing Co., Inc., New York. Chapter 2.

HUTT, F. B., and B. A. RASMUSEN. 1982. *Animal Genetics*, 2nd ed. John Wiley & Sons, Inc., New York. Chapters 2 and 3.

STANSFIELD, W. D. 1983. *Theory and Problems of Genetics*, 2nd ed. McGraw-Hill Book Company, New York. Chapter 2.

THREE

Two or More Pairs of Genes

In Chapter Two we became familiar with the way in which a single pair of alleles behaves over a period of two or three generations. We shall see in this chapter that two or more pairs of genes and the traits they control will usually show up in succeeding generations in the same proportions as if they were acting alone. The apparent exceptions to this statement will be studied in several other chapters. In this chapter we will make as much use as possible of the principles set forth in Chapters One and Two.

3.1 CROSSING THE PUREBREDS

To illustrate how two sets of genes and independent characters segregate, let us consider a cross between cattle of the Hereford and Angus breeds. Angus cattle express the dominant black color (B_-), while Herefords express the recessive red (bb). The Hereford spotting pattern, including the well-known white face pattern, is controlled by a dominant gene. The absence of the pattern, solid color, is the result of a recessive allele. The letters H and h will be used to represent the dominant and recessive alleles respectively, for simplicity. A more detailed description of these and other colors and patterns will be presented in Chapter Eight using different symbols.

When members of these breeds are mated together, the two sets of characters tend to segregate in all possible combinations expected due to chance. Among the F_2 generation we will expect to see black cattle with white faces and solid red cattle, as well as solid black and white-faced red cattle common to the purebreds. This principle of segregation and recombination of traits is known as the *second law of genetics*. It can also be referred to as *Mendel's second law*, the *law of independent assortment*, or the *law of random assortment of characters*.

Crosses between solid black cattle and white-faced red cattle that are homozygous for these traits produce the so-called familiar "black baldies," calves that express the dominant black of the Angus and the dominant Hereford pattern. With respect to these two independent sets of characters, the F_1 calves are heterozygous for two sets of alleles and are called *dihybrids*. The parental or P_1 cross can be illustrated as follows:

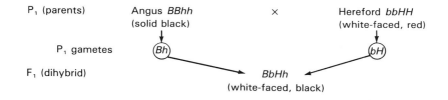

3.2 THE DIHYBRID CROSS

A cross between two of the F_1 black, white-faced animals from the P_1 cross above constitutes a dihybrid cross. There are at least two methods that can be used to solve this cross. One method, the Punnett square method, makes use of a checkerboard-like format. The other is an algebraic method that makes use of the principles established in Chapter Two.

3.2.1 The Punnett Square Method

The *Punnett square method* of solving genetic crosses better illustrates segregation of genes in the gametes and their recombination at fertilization than does the bracket or algebraic method. The checkerboard is merely a geometrical device to help in putting together all possible kinds of gametes from each parent in the cross. It is a good learning exercise for a better understanding of segregation and recombination of genes, but is more time-consuming than the algebraic method.

The first step in the process is to determine all possible kinds of gametes that can be produced by the two animals involved in the mating. From the dihybrid *BbHh*, we know that since there are two pairs of genes involved, each gamete must have only one gene from each locus, that is, a gene from the *B*-locus and one from the *H*-locus. Since the genes at the two loci are expected to segregate independently, there are two possibilities from each locus, or a total of four possible combinations. These combinations are *BH, Bh, bH*, and *bh*.

Since the genotypes of the two participants in a dihybrid cross are the same (*BbHh* × *BbHh*), each should produce gametes of the same kinds and in the same proportions. After the number and kinds of gametes have been determined, a chart is prepared with the gametes from one individual posted across the top of the chart and those of the other posted down the left side of the chart as shown in Figure 3.1. Intersecting rows and columns give the chart the appearance of a checkerboard.

The theoretical or expected genotypes and genotypic ratio of the dihybrid cross are

1 *BBHH*	2 *BbHH*	1 *bbHH*
2 *BBHh*	4 *BbHh*	2 *bbHh*
1 *BBhh*	2 *Bbhh*	1 *bbhh*

Although the gene symbols will vary from one problem to the next, the

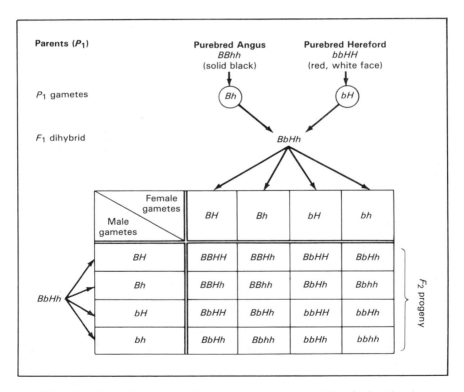

Figure 3.1   F₁ and F₂ progeny of matings between Angus and Hereford cattle using the Punnett square method.

expected ratios will be the same for all dihybrid crosses, regardless of the type of gene action involved.

A careful observation of the $1:2:1:2:4:2:1:2:1$ ratio will reveal several interesting characteristics that should prove helpful later in solving other dihybrid crosses. Notice that the four genotypes associated with the 2's in the ratio each consist of one pair of homozygous and one pair of heterozygous genes: *BBHh*, *bbHh*, *BbHH*, and *Bbhh*. The 1's are associated with genotypes that are homozygous for each of two pairs of genes: *BBHH*, *BBhh*, *bbHH*, and *bbhh*. Finally, the one genotype associated with the 4 is heterozygous at both loci—a dihybrid.

Once the genotypes of the progeny from the dihybrid cross are determined, the phenotypes can be determined. A count of the phenotypes in Figure 3.1 reveals a phenotypic ratio of

> 9 white-faced black (two dominant traits)
> 3 white-faced red (one dominant and one recessive trait)
> 3 solid black (one recessive and one dominant trait)
> 1 solid red (two recessive traits)

The $9:3:3:1$ ratio is the classical phenotypic ratio of every dihybrid cross involving two independent pairs of genes both controlled by dominance and recessiveness.

Of the 16 possible combinations of genotypes, nine of them are associated with the two dominant phenotypes. But among the nine possible combinations, there are only four different ones: 1 *BBHH*, 2 *BBHh*, 2 *BbHH*, and 4 *BbHh*. Note that only one of these nine genotypes is homozygous for both pair of genes. Only one of the 16 possible combinations expresses the two recessive traits, and it is completely homozygous (*bbhh*).

Of the six remaining combinations, three express one dominant trait and one recessive trait, while the other three express the alternate dominant and recessive traits. Among the three individuals in each of these two groups, there is a $2:1$ ratio of genotypes for each dominant trait. For example, among the three solid red calves, there are two *bbHh* and one *bbHH* genotypes. Similarly, among the three white-faced black calves, there are two *Bbhh* and one *BBhh* genotypes.

There are 45 different possible crosses involving two pairs of genes and two alleles each. The dihybrid cross is only one of those possible crosses. The square method can be used to solve any of them. The number of columns and rows, and thus the number of cells, in the chart will be determined by the number and kinds of gametes that each genotype of parent can produce. Figure 3.2 shows several examples of two-pair crosses that are not dihybrid crosses. Notice that none of the examples has as many as 16 cells in the chart. In fact, the dihybrid cross is the only one of the many two-pair crosses that requires this many cells to solve.

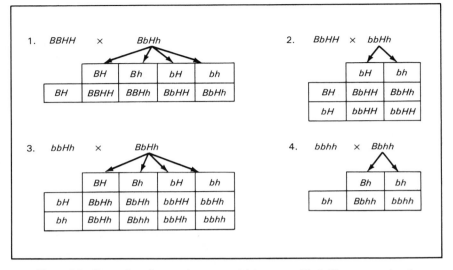

Figure 3.2 Examples of two-pair crosses which are not dihybrid crosses, using the square method of solution.

3.2.2 The Algebraic or Bracket Method

The *algebraic* or *bracket method* is less time-consuming than the square method and provides fewer opportunities for error. The key factor is to make use of the results of monohybrid and other one-pair crosses. The dihybrid cross can be visualized as two simultaneous monohybrid crosses. One cross deals with the genes for black and red: $Bb \times Bb$. Of the calves produced from crosses among cattle of the heterozygous genotype, three-fourths are expected to be black and one-fourth are expected to be red.

The second monohybrid cross involves the genes for the presence or absence of white markings of the Hereford pattern: $Hh \times Hh$. The product of this cross is also the theoretical 3 : 1 ratio of white face to solid color, respectively. Because the two sets of genes are inherited independently of each other, the two sets of traits are also expected to be distributed among the F_2 generation at random. Three-fourths of the black cattle should have a white face, while one-fourth should be solid black; and three-fourths of the red cattle should have a white face, while one-fourth of them should be solid red. The final results can be obtained by multiplying the two mono-hybrid F_2 ratios together: (3 black : 1 red) × (3 white-faced : 1 solid color) =

9 white-faced black
3 white-faced red
3 solid, black
1 solid, red
_____

16 total possible combinations

The two monohybrid ratios could just as easily be multiplied as fractions: ($\frac{3}{4}$ black : $\frac{1}{4}$ red) $\times$ ($\frac{3}{4}$ white-faced : $\frac{1}{4}$ solid color) =

$\frac{9}{16}$ white-faced black

$\frac{3}{16}$ white-faced red

$\frac{3}{16}$ solid, black

$\frac{1}{16}$ solid, red

The 9 : 3 : 3 : 1 ratio is made up of the numerators of the fractions $\frac{9}{16} : \frac{3}{16} : \frac{3}{16} : \frac{1}{16}$, respectively. The sum of all the numerators is equal to the denominator.

A slight variation of the algebraic method involves the same principle but makes use of brackets:

3 black $\Bigg\langle$ 3 white face $\longrightarrow$ 9 white-faced, black
1 solid $\longrightarrow$ 3 solid, black

1 red $\Bigg\langle$ 3 white face $\longrightarrow$ 3 white-faced, red
1 solid $\longrightarrow$ 1 solid, red

3.3 THE TWO-PAIR TESTCROSS

Another cross that has an important purpose among two-pair crosses, particularly crosses involving two pairs of dominant and recessive genes, is the testcross. The purpose of the testcross is the same regardless of the number of pairs of genes involved—to determine if an individual expressing one or more dominant traits carries recessive genes or not ($B?H? \times bbhh$).

Assume that a black white-faced bull is chosen at random and we would like to determine his actual genotype; he can be tested by mating to several solid red cows. The bull could be any one of four possible genotypes: *BBHH, BBHh, BbHH,* or *BbHh.* The phenotypes of the testcross progeny would indicate which is the most probable genotype. Figure 3.3 shows the results of the four possible crosses.

If the number of progeny is large enough to allow all possible combinations to occur, those very results would indicate the probable genotype

1. BbHh × bbhh	2. BbHH × bbhh
1 *BbHh,* white-faced, black 1 *Bbhh,* solid black 1 *bbHh,* white-faced, red 1 *bbhh,* solid red	1 *BbHH,* white-faced, black 1 *bbHh,* white-faced, red
3. BBHH × bbhh	4. BBHh × bbhh
all *BbHh,* white-faced, black	1 *BbHh,* white-faced, black 1 *Bbhh,* solid, black

Figure 3.3 Four possible sets of genotypes and phenotypes expected from a dihybrid testcross.

of the black, white-faced parent. For example, if no red calves were produced among five to 10 calves, this would indicate that the black parent is very likely homozygous (*BB*). Further, if among the 5 to 10 calves, all had white faces, this would indicate that the white-faced parent was probably homozygous (*HH*).

3.4 MODIFICATIONS OF THE CLASSICAL DIHYBRID RATIO

It was stated earlier that genetics is more than dominance and recessiveness and 3 : 1 ratios. As far as two-pair crosses are concerned, it is also more than dominance and recessiveness and 9 : 3 : 3 : 1 ratios. Two kinds of gene action were discussed in Chapter Two as to how they modified the classical monohybrid ratio—these were codominance and the effects of lethal genes. We shall now see how these kinds of gene action affect two-pair ratios, particularly the classical 9 : 3 : 3 : 1 ratio that is expected from a dihybrid cross involving two pairs of dominant and recessive genes. In the next few chapters, several other kinds of gene action are presented that will modify these ratios even further.

3.4.1 Effect of Codominant Alleles

Black in some breeds of swine is controlled by the homozygous state of one of two codominant alleles (*AA*), the second of which produces red when in the homozygous state (*aa*). The heterozygous state of the two alleles produces a red-and-black spotted pattern (*Aa*). The type of inheritance is very much the same as that which controls red, roan, and white in Shorthorn cattle. The phenotypic ratio of the monohybrid cross (*Aa* × *Aa*) is $\frac{1}{4}$ black : $\frac{1}{2}$ spotted : $\frac{1}{4}$ red.

Erect ears in swine are thought to be controlled by a dominant gene

($E\_$); drooping ears are due to the presence of the recessive allele in the homozygous state (ee). The monohybrid cross ($Ee \times Ee$) would produce the classical ratio of 3 : 1 for erect and drooping ears, respectively.

The dihybrid ($AaEe$) can be produced by crossing purebred Berkshires, assumed to be homozygous for black and erect ears, with purebred Durocs, which are red with drooping ears. The F_1 would have erect ears and be red and black in color.

The ratio of F_2 phenotypes can be determined by multiplying the monohybrid ratios from the two preceding paragraphs: ($\frac{1}{4}$ black : $\frac{1}{2}$ spotted : $\frac{1}{4}$ red) $\times$ ($\frac{3}{4}$ erect ears : $\frac{1}{4}$ drooping ears) =

$\frac{3}{16}$ black with erect ears

$\frac{3}{8}$ spotted with erect ears

$\frac{3}{16}$ red with erect ears

$\frac{1}{16}$ black with drooping ears

$\frac{1}{8}$ spotted with drooping ears

$\frac{1}{16}$ red with drooping ears

In summary, the phenotypic ratio of a dihybrid cross involving one pair of alleles showing dominance and recessiveness and one pair of codominant alleles is 3 : 6 : 3 : 1 : 2 : 1.

Combining the results of two sets of codominant alleles would be expected to produce a phenotypic ratio the same as the genotypic ratio for the same kind of cross. The feather colors black, blue, and white in Andalusian chickens are controlled by a pair of codominant alleles such that black and white are homozygous, and blue is heterozygous. The phenotypic ratio of the monohybrid cross would be 1 black : 2 blue : 1 white.

A second set of traits in chickens controlled by a pair of codominant alleles involves a feather condition called frizzling. The presence of normal feathers is due to the homozygous state of one allele (ff). The homozygote for the frizzle gene (FF) produces a bird with feathers that curl forward, producing an extremely "wooly" appearance. The abnormality results in a loss of the insulating value of the feathers as well as a loss of the ability to fly. The heterozygote (Ff) results in a medium frizzling of the feathers that is desired by many fanciers who exhibit them at poultry shows.

Assume that blue, medium-frizzled hens and cocks were mated and produced 76 chicks that eventually showed the following phenotypes and numbers:

   4 black with normal feathers
   8 blue with normal feathers
   5 white with normal feathers
   9 black with medium frizzling
  19 blue with medium frizzling
  12 white with medium frizzling

$\quad$ 4 black with extreme frizzling
10 blue with extreme frizzling
$\quad$ 5 white with extreme frizzling
$\overline{76}$ total

Even though the ratio shown here is not a perfect $1:2:1:2:4:2:1:2:1$, the deviations are not greater than what might be expected due to chance.

3.4.2 Effect of Lethal Genes

$\quad$ The way by which a lethal gene that expresses a lack of dominance or recessiveness with respect to its allele modifies the results of the monohybrid cross was presented in Chapter Two. The creeper gene in chickens is an example of such a gene. When homozygous, this gene causes the death of the embryo before hatching. In the heterozygous state, the creeper gene and its normal allele interact to produce the creeper trait, characterized by abnormal development of the legs such that the bird appears to "creep." The phenotypic ratio of a monohybrid cross involving the creeper trait is $\frac{2}{3}$ creepers : $\frac{1}{3}$ normals.

$\quad$ Black feather color in chickens is controlled by a dominant gene, while red feather color is the result of the homozygous state of the recessive allele. Heterozygous black creepers mated among themselves constitute a dihybrid cross. The phenotypic results of this cross can be determined quickly by algebraically multiplying the phenotypic results of the two monohybrid crosses: ($\frac{3}{4}$ black : $\frac{1}{4}$ red) $\times$ ($\frac{2}{3}$ creeper : $\frac{1}{3}$ normal) $=$ ($\frac{6}{12}$ black creepers : $\frac{3}{12}$ black normals : $\frac{2}{12}$ red creepers : $\frac{1}{12}$ red normals). If these fractions are reduced, the ratio could be expressed as $\frac{1}{2}:\frac{1}{4}:\frac{1}{6}:\frac{1}{12}$. The other common way of expressing the ratio is to use only the numerators of the fractions all having the same denominators: $6:3:2:1$.

$\quad$ Other examples of combinations of traits controlled by lethal genes and other types of gene action can be given (see Table 3.1). In each case, the lethal gene has the effect of reducing the total number of progeny compared to that expected with normal ratios.

$\quad$ A normal trait controlled by codominance expressed in conjunction with another trait controlled by codominant alleles, one of which is lethal, would produce these results: $(1:2:1) \times (2:1) = 2:4:2:1:2:1$. If we could find two sets of traits both consisting of normal and lethal alleles, the results would be $(2:1) \times (2:1) = 4:2:2:1$.

3.5 THREE-PAIR CROSSES

In the same way that principles associated with one-pair crosses are used to solve two-pair crosses, they can also be used in the solution of crosses involving three or more pairs of genes.

TABLE 3.1   Allelic Relationships in Dihybrid Parents and Their Progeny

Allelic Relationships of Dihybrid Parents		Expected Phenotypic Ratios of Progeny
First Locus	*Second Locus*	
Dominant–recessive	Dominant–recessive	9 : 3 : 3 : 1
Dominant–recessive	Codominant	3 : 6 : 3 : 1 : 2 : 1
Dominant–recessive	Codominant lethal	6 : 3 : 2 : 1
Codominant	Codominant	1 : 2 : 1 : 2 : 4 : 2 : 1 : 2 : 1
Codominant	Codominant lethal	2 : 4 : 2 : 1 : 2 : 1
Codominant lethal	Codominant lethal	4 : 2 : 2 : 1

To illustrate, let us consider three sets of dominant and recessive genes. In addition to the two sets of traits associated with purebred Herefords and Angus cattle that were used in the illustration at the beginning of this chapter, American Herefords express the horned trait, which is the result of the homozygous state of a recessive gene (*pp*). Polledness, the absence of horns, is controlled by the dominant allele, *P_*, and is characteristic of purebred Angus cattle.

Crosses between purebred Angus and American Herefords would produce the familiar black "baldies" as illustrated earlier, but in addition, the F₁ progeny would also be heterozygous for the polled trait:

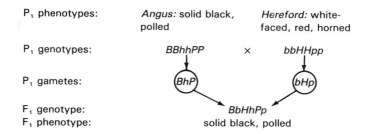

P₁ phenotypes:	*Angus:* solid black, polled	*Hereford:* white-faced, red, horned
P₁ genotypes:	*BBhhPP* ×	*bbHHpp*
P₁ gametes:	*BhP*	*bHp*
F₁ genotype:	*BbHhPp*	
F₁ phenotype:	solid black, polled	

The solution of the trihybrid cross by means of the Punnett square method would be quite a bit more complex and time-consuming than the dihybrid cross. The trihybrid would produce eight different kinds of gametes, which means that the chart used to obtain all possible combinations of F₂ genotypes would have eight rows, eight columns, and 64 individual cells or squares.

The simpler procedure is to use the algebraic method of combining the three monohybrid ratios: (3 black : 1 red) × (3 white-faced : 1 solid color) × (3 polled : 1 horned) =

27 white-faced, black, polled
9 white-faced, black, horned

     9 solid black, polled
     9 white-faced, red, polled
     3 solid black, horned
     3 white-faced, red, horned
     3 solid red, polled
    <u> 1</u> solid red, horned
    64 total combinations

STUDY QUESTIONS AND EXERCISES

3.1. How does a dihybrid cross differ from any other cross involving two pairs of genes?

3.2. Explain the second law of genetics and compare it to the first law.

3.3. Compare the advantages of the square method and the algebraic method of solving two-pair crosses.

3.4. To what extent are genotypic ratios of genetic crosses affected by the kind of gene action involved? Explain.

3.5. Of the many different examples of two-pair crosses, how many of them are dihybrid crosses? Explain.

3.6. What is the function of the two-pair testcross?

3.7. How does the involvement of traits controlled by codominant alleles affect the number of different phenotypes produced by the dihybrid cross when compared to the classical 9 : 3 : 3 : 1 ratio?

3.8. How does the involvement of lethal genes in a dihybrid cross affect the total possible number of individuals produced?

3.9. Show why the Punnett square method of solving a trihybrid cross is so much more complex than using the algebraic method.

PROBLEMS

3.1. List the minimum number of all possible gametes that can be formed from the following genotypes: **(a)** *AABbcc*; **(b)** *AaBb*; **(c)** *AabbCCdd*; **(d)** *aaBbCc-DdEE*; **(e)** *AAbbCCdd*.

3.2. How many cells (individual squares) would be required to solve each of the following crosses using the Punnett square method? **(a)** *BBccDd* × *bbCcDd*; **(b)** *AaBB* × *aaBb*; **(c)** *AAbbcc* × *aaBbcc*: **(d)** *DDeeFF* × *ddEEff*; **(e)** *AaBbCc* × *AaBbCc*; **(f)** *AabbCc* × *aabbcc*.

3.3. Determine the expected phenotypic ratios for each of the following combinations of one-pair ratios: **(a)** (3 : 1) × (2 : 1); **(b)** (1 : 1) × (1 : 1); **(c)** (3 : 1) × (3 : 1) × (3 : 1); **(d)** (1 : 2 : 1) × (3 : 1); **(e)** (1 : 2 : 1) × (1 : 2 : 1); **(f)** (3 : 1) × (2 : 1) × (1 : 1).

3.4. How many different kinds of genotypes are possible among the progeny of a dihybrid cross?

3.5. Black color in Hampshire (*B_*) swine is dominant to the red (*bb*), as expressed in Duroc swine. The white belt (*H_*) of Hampshires is also dominant to the absence of the belt (*hh*).

   (a) Assuming homozygosity in the purebreds, what would be the results of crossing purebred Hampshires with purebred Durocs?

   (b) If the F_1 progeny of this cross were mated among themselves, indicate all possible genotypes that would be expected, using the Punnett square method.

   (c) Show the formula that would be used to determine the phenotypes and phenotypic ratio of the dihybrid cross, and show these results.

   (d) Assume that a testcross of a black-belted sow produced a litter of seven pigs, three of which were black belted and four of which were red belted. What is the probable genotype of the sow?

   (e) Of the four combinations of traits possible, which would be the easiest to establish in a pure-breeding group of swine? Why?

   (f) Why do Hampshire swine occasionally produce a pig without the white belt?

   (g) Choose only the belted red hogs from the F_2 progeny in part (c) and allow them to mate together at random. What genotypic and phenotypic results would be expected among the progeny?

3.6. In poultry, the pea comb trait (*P–*) is dominant to the single-comb trait (*pp*). Black feathers are controlled by the homozygous state of a gene (*B*) that is codominant to its allele for white (*b*). The heterozygote (*Bb*) produces blue feathers. List the genotypes and phenotypes, including ratios, expected in the progeny from each of these crosses: **(a)** *BBpp* × *BbPp*; **(b)** *Bbpp* × *bbPp*; **(c)** *BbPp* × *Bbpp*; **(d)** *bbPP* × *BBpp*; **(e)** *bbPp* × *bbPp*.

3.7. A white hen with a pea comb produced a brood of 13 chicks. When their combs and feathers developed, two birds were blue with pea combs, four had white feathers and pea combs, two had white feathers and single combs, and one had blue feathers and a single comb.

   (a) What is the probable genotype and phenotype of the cock that sired these chicks?

   (b) What is the genotype of the hen?

   (c) How many of the four white pea-combed progeny would be expected to be heterozygous for the pea comb gene?

   (d) Each of the four progeny with white feathers and pea combs were eventually mated in a testcross; each produced at least one progeny with the single-comb trait. How many of those original four progeny would you predict as being heterozygous?

3.8. The short-hair trait in rabbits is controlled by a dominant gene *S; long hair is* controlled by the recessive allele *s*. An abnormality of the white blood cells called the Pelger anomaly is caused by the heterozygous state of two alleles (*Pp*). Rabbits with normal cells are homozygous (*pp*). The homozygous state of the *P* allele causes abnormal development of the skeleton in the embryo usually causing death before or very soon after birth.

   (a) Suppose that heterozygous short-haired Pelgers are mated together. What

phenotypic expectations would be produced among several litters from
such matings at weaning time?

(b) Would it be possible to produce a litter in which all the progeny would be
expected to be Pelgers? Why?

(c) Assume that the average litter size at weaning when normal rabbits are
mated together is four; what would you predict as the expected litter size
when Pelgers are mated to Pelgers?

(d) What would you predict as to the average litter size at weaning when Pelgers and normals are mated together?

REFERENCES

BURNS, G. W. 1980. *The Science of Genetics,* 4th ed. Macmillan Publishing Co.,
Inc., New York. Chapter 4.

HUTT, F. B., and B. A. RASMUSEN. 1982. *Animal Genetics,* 2nd ed. John Wiley &
Sons, Inc., New York. Chapter 4.

STANSFIELD, W. D. 1983. *Theory and Problems of Genetics*, 2nd ed. McGraw-Hill
Book Company, New York. Chapter 3.

FOUR

Probabilities
and Chi-Square

Probability deals with our degree of sureness as to whether or not a certain event will occur. A knowledge and understanding of the laws of probability are fundamental to a more complete understanding of genetics. The classical genotypic and phenotypic ratios presented in Chapters Two and Three were determined by making use of laws of probability. Some of the principles of probabilities and their applications are presented in this chapter. Included among these applications are: (1) predicting genotypic and phenotypic results of various crosses (as in Chapters Two and Three), (2) determining how well the observed results of any cross fit the predicted or expected results of that cross, (3) drawing conclusions as to whether the genotype of an individual does or does not include a certain recessive gene, and (4) predicting the number of individuals among litters and other groups of progeny that possess alternative sexes or phenotypes.

4.1 PROBABILITIES OF NONGENETIC EVENTS

Probabilities have become an intimate part of our daily conversation and activities. There is very little about which we are absolutely sure; we would rather use terms such as *probably, likelihood, chances,* or *odds* to discuss the occurrence of various events. Weather reports and discussions of outcomes of football and basketball games are filled with probability terms. If we make an appointment to meet someone, we seldom say "I will be there at such-and-such a time"; we more often will say "I should arrive there about noon" or "I will probably see you around 12 o'clock."

The laws of probability can be explained and illustrated using any chance or random event, such as tossing a coin, drawing a card from a deck of playing cards, or rolling a die. Laws of probability have long been an important part of games of chance.

A coin tossed into the air and allowed to come to rest is likely to land as either "heads" or "tails." One of these events is as likely to occur as the other. It can therefore be stated that the probability of a head in one toss of a coin is $\frac{1}{2}$ or 0.5 or 50 percent. The probability of obtaining a tail is also $\frac{1}{2}$. If we were to toss a single coin 100 times, for example, we would expect heads to appear about as often as tails. Similar results would be expected if we were to toss 100 different coins simultaneously. If we were to toss four coins, however, we would not always expect two to appear as heads and two to appear as tails. It would not be considered unusual to see three heads and one tail or three tails and one head. It is possible also to expect four heads or four tails occasionally. The chances of each of the occurrences can be predicted using laws of probability.

Let us use the simpler illustration of tossing two coins, say a penny and a dime. The probability of obtaining a head on the penny is $\frac{1}{2}$; the probability of obtaining a head on the dime is also $\frac{1}{2}$. What is the probability of obtaining a head on both the penny and the dime if they are tossed simultaneously? Whether the penny lands tails up or heads up is totally independent of how the dime will land. These two events are therefore defined as independent events. The probability that two or more independent events will occur simultaneously is the product of probabilities that each event will occur separately. Therefore, the probability that two heads will occur if two coins are tossed simultaneously is $\frac{1}{2} \times \frac{1}{2}$ or $\frac{1}{4}$. The probability that two coins tossed simultaneously will both land with tails up is also $\frac{1}{4}$ ($\frac{1}{2} \times \frac{1}{2}$).

Two coin tosses can result in two heads, two tails, or one tail and one head. Since these are the only three possibilities and the probabilities of two heads or two tails are $\frac{1}{4}$ each, it is easy to see that the probability that one head and one tail will occur from the toss of two coins must be $\frac{1}{2}$. But why is the probability of this combination greater than either of the other two? It can be seen from the following illustration that there are two ways to obtain a head and a tail from the toss of two coins:

First Coin	Second Coin	Probability
Head	Head	$\frac{1}{4}$
Head	Tail	$\frac{1}{4}$
Tail	Head	$\frac{1}{4}$
Tail	Tail	$\frac{1}{4}$

Each of the two ways of obtaining one head and one tail has a probability of $\frac{1}{4}$ of occurring. If a head is obtained on the first coin and a tail is obtained on the second, the opposite combination cannot be obtained at the same time. Since the occurrence of one of these events excludes the occurrence of the other, they are said to be *mutually exclusive*. The rule which governs the combining of such events states that the probability that one or another of two or more mutually exclusive events will occur is the sum of their individual probabilities. For example, if two coins are tossed, there are two ways to obtain one head and one tail. A head can occur on the first and a tail on the second (the probability is $\frac{1}{2} \times \frac{1}{2} = \frac{1}{4}$), or a tail on the first and a head on the second (this probability is also $\frac{1}{2} \times \frac{1}{2} = \frac{1}{4}$). However, one combination of events excludes the other, so if the order of heads and tails is not important, the probability of a head and a tail in either order is $\frac{1}{4} + \frac{1}{4} = \frac{1}{2}$.

Compare the rule for combining probabilities of mutually exclusive events with that for combining independent events. Note that the differences in the wording of the two rules relate to the ways in which the probabilities of the two kinds of events are combined. The probabilities of *independent* events are *multiplied* when combined; the probabilities of *mutually exclusive* events are *added* when combined.

4.2 PROBABILITIES OF GENETIC EVENTS

4.2.1 Predicting Genetic Ratios

The genetic law of segregation and recombination is based on laws of probability. Genes are expected to segregate in the gametes during meiosis in all possible combinations due to chance. An individual of the genotype AA will produce gametes containing only the gene A. Stating it another way, the probability that an individual of the genotype AA will produce a gamete containing the gene A is 1.0 (or 100 percent). The probability that the same individual will produce a gamete containing the gene a is zero if we disregard the possible occurrence of a mutation. These are the probabilities that are expected due to chance. Similarly, the probability that an individual of the genotype aa will produce a gamete containing the a gene is 1.0; the probability that the same individual will produce a gamete containing the A gene is zero.

The probabilities associated with the formation of gametes by an individual of the genotype Aa are very similar to those asssociated with the tossing of a coin. A coin can produce either heads or tails when tossed; the heterozygote Aa can produce gametes containing either the gene A or its allele a, and with equal probabilities. The probability that an individual of the genotype Aa will produce a gamete containing A is $\frac{1}{2}$; the probability that this individual will produce a gamete containing the gene a is also $\frac{1}{2}$. These are the results expected due to chance.

When crosses are performed among individuals of various genotypes, the genes in the gametes are expected to recombine in all possible combinations due to chance. The several crosses involving two alleles at one locus were illustrated in Chapter Two. To illustrate how the laws of probability apply, let us consider only one of the six basic crosses, the monohybrid cross $Aa \times Aa$. We read in the preceding paragraph that each heterozygote is expected to produce two kinds of gametes, one containing the gene A and a second containing the gene a, each occurring with the probability $\frac{1}{2}$.

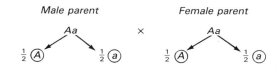

(Kinds of gametes expected and the probabilities of their occurrence)

Whether a male gamete contains the gene A or a is independent of whether a female gamete contains the gene A or a. Therefore, the probability that a male gamete A will unite with a female gamete A (assuming that fertilization does occur) is $\frac{1}{2} \times \frac{1}{2}$ or $\frac{1}{4}$. Similarly, the probability that two gametes containing the genes a will unite is also $\frac{1}{4}$. These are the same probabilities as those expected for obtaining two heads or two tails from the tossing of two coins.

The production of a heterozygote from a monohybrid cross can occur in one of two ways. A male gamete containing the gene A can unite with a female gamete containing the gene a, or a male gamete containing the gene a can unite with a female gamete containing the gene A. Each occurrence has a probability of $\frac{1}{4}$. However, the occurrence of one eventuality excludes the occurrence of the other, so the probability that a heterozygote will result from a monohybrid cross is $\frac{1}{4} + \frac{1}{4}$ or $\frac{1}{2}$. The theoretical genotypic ratio of $1:2:1$ obtained from a monohybrid cross represents the numerators of fractions associated with the probabilities of the three expected genotypes: $\frac{1}{4} AA : \frac{2}{4} Aa : \frac{1}{4} aa$.

The laws of probability can be applied to crosses involving two or more pairs of genes. Consider a dihybrid cross involving two sets of dominant and recessive genes in swine. Black in some breeds of swine is due to a dominant gene B, while red is due to the recessive allele b. The white belt observed in Hampshires is controlled by a dominant gene H and the absence of the belt is controlled by the recessive allele h. What is the probability that one pig produced from two dihybrid parents ($BbHh \times BbHh$) will be red and have the white belt? Each hybrid parent can produce four kinds of gametes: *BH*, *Bh*, *bH*, and *bh*, each occurring with a probability of $\frac{1}{4}$. A belted red pig can be produced in any of three ways:

Male gamete		Female gamete		Genotype of the pig
$\frac{1}{4}$ *bH*	x	$\frac{1}{4}$ *bH*	=	$\frac{1}{16}$ *bbHH*
$\frac{1}{4}$ *bH*	x	$\frac{1}{4}$ *bh*	=	$\frac{1}{16}$ *bbHh*
$\frac{1}{4}$ *bh*	x	$\frac{1}{4}$ *bH*	=	$\frac{1}{16}$ *bbHh*

Since each type of gamete produced by the parents occurs with a probability of $\frac{1}{4}$, the combination of any two gametes will occur with a probability of $\frac{1}{16}$ ($\frac{1}{4} \times \frac{1}{4}$) because we are combining independent events. The three separate ways of obtaining a genotype that will result in a belted red pig are mutually exclusive; therefore, the probability that this phenotype will occur from the dihybrid cross is determined by adding the three separate probabilities ($\frac{1}{16} + \frac{1}{16} + \frac{1}{16} = \frac{3}{16}$). If you were to determine all possible phenotypes that could be produced from this hybrid cross, the theoretical results would be $\frac{9}{16}$ belted black pigs, $\frac{3}{16}$ belted red pigs, $\frac{3}{16}$ nonbelted black pigs, and $\frac{1}{16}$ nonbelted red pigs. The results of the two procedures with respect to the probability of obtaining a belted red pig are the same.

A second approach to solving the question would be to combine the expected results of the two monohybrid crosses contained within the dihybrid cross. This principle was explained in Chapter Three. The probability that the mating of heterozygous black swine will produce a red pig is $\frac{1}{4}$. Recall that the phenotypic ratio of a monohybrid cross involving dominant and recessive genes is $\frac{3}{4}$ dominant and $\frac{1}{4}$ recessive. The probability that the mating of heterozygous belted swine will produce a belted pig is $\frac{3}{4}$. The occurrence of the belt is independent of the color of the pigs, so the probability that the mating of dihybrid belted black swine will produce a belted red pig can be calculated by multiplying the probabilities $\frac{1}{4}$ (for red) and $\frac{3}{4}$ (for belted), the result being $\frac{3}{16}$ as obtained by the first approach.

4.2.2 Determining the Presence of a Recessive Gene

As noted in Chapters Two and Three, one of the problems with traits controlled by genes with complete dominance is that the genotype may be either homozygous or heterozygous. There are occasions when it is important to know whether the recessive gene is present or not. This is particularly true if the recessive gene controls a trait that is lethal or detrimental, or otherwise objectionable. The laws of probability can be used to help make this determination.

Black in Angus and Holstein cattle and in Hampshire swine is a dominant trait. It would be objectionable in purebred herds of black animals if some red ones were to be produced, since black is one of the identifying characteristics of these breeds. Consider the Hampshire breed of swine, for example. Assume that a breeder has purchased a young boar but is not sure whether he is homozygous (*BB*) or heterozygous (*Bb*) for black. One way to determine his genotype would be to perform the testcross; that is, mate him to some red sows. If he is homozygous, he would never be expected to sire any red pigs—*BB* (black boar) × *bb* (red sows) would produce all black (*Bb*) pigs. If he were heterozygous, one-half of his progeny would be expected to be red when mated to red sows, $Bb \times bb = \frac{1}{2}$ *Bb* (black) pigs and $\frac{1}{2}$ *bb* (red) pigs. If he were to sire one or more red pigs, you would know immediately that his genotype was heterozygous. If he did not sire any red pigs, however, how sure could we be that he was homozygous? He might be heterozygous and due to chance, no gametes bearing the red gene were involved in the fertilization process. This would be very much like the situation of obtaining all tails (or all heads) from the continued tossing of a coin.

If the boar were heterozygous, let us calculate the probabilities of his siring any number of black pigs when mated to red sows, but not producing any red pigs. The probability of one black pig being produced is $\frac{1}{2}$. The probability of his siring any number of black pigs would be $(\frac{1}{2})^n$, where n is the number of pigs sired. The question the breeder must answer for himself is: How many black pigs must the boar sire without producing any red ones before it is decided that he is probably homozygous? It is generally considered reasonable to expect a level of confidence of 95 percent or greater under circumstances such as this. In other words, if the breeder were to decide that the boar was homozygous and his chances of being wrong were only 1 in 20 (5 percent), the level of confidence that his decision was correct would be 95 percent (100 percent minus 5 percent).

Note in Table 4.1 that the probability that five offspring would be produced from mating a hererozygote to recessive individuals (testcross) without producing any recessive offspring is 0.03125 (or $3\frac{1}{8}$ percent). Thus, if it is decided that since no red pigs were produced among the five, the boar is not heterozygous, the confidence level of that decision would be approximately 97 percent. If the breeder wanted a greater level of confidence that the boar were homozygous, several more offspring could be produced. One litter producing just seven black pigs would provide the breeder with a level of confidence of greater than 99 percent if it were decided that the boar was homozygous. Ten black offspring from the testcross of a heterozygote would be expected with a probability of less than 1 in 1000. This would be similar to tossing a coin 10 times and producing 10 heads.

With some traits, it may not be practical to perform a testcross. The nature of the recessive trait may make it impractical or even impossible for individuals bearing that trait to be used for breeding purposes. A type of

TABLE 4.1   Probabilities of Obtaining No Recessive Offspring
in Various Kinds of Matings with Heterozygotes

Number of Progeny, n	Testcross, $(\frac{1}{2})^n$	Mating to Known Heterozygotes, $(\frac{3}{4})^n$	Mating to Daughters of Known Heterozygotes, $(\frac{7}{8})^n$
1	0.5	0.75	0.875
2	0.25	0.5625	0.765625
3	0.125	0.421875	0.669921
4	0.0625	0.316406	0.586182
5	0.03125	0.237305	0.512909
6	0.015125	0.177979	0.448795
7	0.007813	0.133484	0.392700
8	0.003906	0.100113	0.343609
9	0.001903	0.075085	0.300657
10	0.000976	0.056313	0.263076
11	0.000488	0.042235	0.230191
17	0.00000763	0.007517	0.103309
23	0.000000119	0.001338	0.046365
35	0.000000000029	0.000042	0.009339

recessive dwarfism in cattle called "snorter" dwarfism is a good example. Dwarf cattle exhibit a greater mortality rate and lower levels of fertility; dwarf cows have difficulty having calves naturally; and such cattle have essentially no commercial value. About the only importance they might have would be for research purposes in studying the genetics and physiology of the trait.

A bull of normal size might be suspected of carrying the recessive dwarf gene (Dd). Since a testcross to dwarf cows (dd) would not be practical, matings could be made to cows with normal phenotypes, but known to be carriers of the dwarf gene. Normal cows that had borne dwarf calves would be in this category. Approximately twice the number of progeny would be needed from this kind of mating as compared to the testcross at any given level of significance. The probability that a normal calf would be produced from mating a heterozygous bull to heterozygous cows is 0.75. To obtain a probability of less than 0.05, for example, 11 normal progeny would be required, with no dwarfs being produced (see Table 4.1). Recall that only five progeny were required from the testcross to obtain this level of significance.

In the event that sufficient known heterozygotes were not available to perform the test, daughters of known heterozygotes could be used. Assuming that known heterozygous cows were mated to homozygous normal bulls, one-half the daughters would also be heterozygous. If a heterozygous bull were mated to a group of daughters of known heterozygotes, the probability that a single normal calf would be produced is $\frac{7}{8}$, or 0.875. This is an average of the probabilities of producing normal progeny from the two types of cows in the group. One-half the group of daughters from known heterozygotes would be expected to be homozygous normal (DD); the other one-half would be expected to be heterozygous. Mating a heterozygous bull to the homozygous daughters would produce all normal calves (probability = 1.00). Mating the bull to the heterozygous females would produce normal calves with an expected probability of 0.75. Since the heterozygous female is just as likely to be chosen in the mating process as is the homozygous female, the two probabilities can be averaged. To declare a bull homozygous with a 95 percent level of confidence when mating him to daughters of known heterozygotes, 23 normal progeny would be required with the production of no dwarves. Note in Table 4.1 the probability that 23 normal progeny would be produced from the mating of heterozygotes to daughters of known heterozygotes is $(\frac{7}{8})^{23}$, or approximately 0.046 (4.6 percent). Based on these results, the bull being tested could be declared homozygous with a confidence level of 95.4 percent (100 − 4.6) that the decision is correct.

4.3 THE BINOMIAL EXPRESSION

For any event that has two alternative outcomes (such as heads or tails from the toss of a coin, the sex of a newborn calf, or the type of gamete produced by a heterozygote), the binomial expression can be used to predict the various combinations of alternatives when two or more of those events occur simultaneously. For example, we might want to know how often to expect 2 boars and 6 gilts in a litter of 8 pigs, or what the probability would be of obtaining at least 8 heads from 10 tosses of a coin.

4.3.1 Characteristics of the Binomial

The general formula for the binomial expression is $(p + q)^n$, where p and q are the probabilities of the two alternative outcomes of the event (heads or tails, male or female, dominant or recessive, etc.), and n is the number of offspring. Once the value of n is determined, the binomial must be expanded to that power. Table 4.2 shows the expansion of the binomial through the ninth power. A knowledge of a few characteristics of the expanded binomial should simplify its use:

1. The number of terms in the expansion is one more than the power of the expansion ($n + 1$).

TABLE 4.2   Expansion of the Binomial

General Formula	Expanded Formula
$(p + q)^2$	$p^2 + 2pq + q^2$
$(p + q)^3$	$p^3 + 3p^2q + 3pq^2 + q^3$
$(p + q)^4$	$p^4 + 4p^3q + 6p^2q^2 + 4pq^3 + q^4$
$(p + q)^5$	$p^5 + 5p^4q + 10p^3q^2 + 10p^2q^3 + 5pq^4 + q^5$
$(p + q)^6$	$p^6 + 6p^5q + 15p^4q^2 + 20p^3q^3 + 15p^2q^4 + 6pq^5 + q^6$
$(p + q)^7$	$p^7 + 7p^6q + 21p^5q^2 + 35p^4q^3 + 35p^3q^4 + 21p^2q^5 + 7pq^6$ $+ q^7$
$(p + q)^8$	$p^8 + 8p^7q + 28p^6q^2 + 56p^5q^3 + 70p^4q^4 + 56p^3q^5 + 28p^2q^6$ $+ 8pq^7 + q^8$
$(p + q)^9$	$p^9 + 9p^8q + 36p^7q^2 + 84p^6q^3 + 126p^5q^4 + 126p^4q^5 + 84p^3q^6$ $+ 36p^2q^7 + 9pq^8 + q^9$

2. The exponents in the binomial expansion follow a definite pattern. The exponents of p begin with the power to which the binomial is raised and decrease to zero; the exponents of q begin with zero and increase to the power to which the binomial is raised.

3. The coefficients of the expanded binomial follow a symmetrical pattern. The coefficients of the first and last terms are both 1 (not written); the coefficients of second term and the second-to-last term are equal to the power to which the binomial is raised. The coefficients of the second and all succeeding terms can be determined using the following formula:

$$\frac{\text{coefficient of}}{\text{any term}} = \frac{\text{coefficient of preceding term} \times \text{exponent of } p \text{ in preceding term}}{\text{number indicating position of preceding term in expansion}}$$

To illustrate the use of the formula above, consider the binomial expanded to the seventh power (see Table 4.2) Suppose that we know the third term in the expansion $(21p^5q^2)$ and want to know the fourth. The coefficient of the preceding term (the third term in this case) is 21; the exponent of p in the third term is 5; therefore, the coefficient of the fourth term will be $(21 \times 5)/3 = 35$.

Coefficients can be determined using the formula $C = n!/a!\,b!$ where $n!$ (called "factorial n") $= n(n - 1)(n - 2) \ldots 1$, n itself is the power of the expansion, a is the exponent of p, and b is the exponent of q. Let us illustrate by calculating the coefficient of the fourth term (the p^4q^3 term) in the binomial expanded to the seventh power. The value of n is 7, $n!$ is 7(6)(5)(4)(3)(2)(1) $= 5040$; $a!\,b! = 4(3)(2)(1) \times (3)(2)(1) = (24)(6) = 144$. So $C = 5040/144 = 35$.

Another method of determining coefficients for any expanded binomial is to use *Pascal's triangle*, where any coefficient is the sum of the nearest two numbers in the row immediately above. For example, 35 is the sum of 20 and 15 in the following chart.

Power											Total
1					1	1					2
2				1	2	1					4
3			1	3	3	1					8
4		1	4	6	4	1					16
5	1	5	10	10	5	1					32
6	1	6	15	20	15	6	1				64
7	1	7	21	35	35	21	7	1			128
8	1	8	28	56	70	56	28	8	1		256
9	1	9	36	84	126	126	84	36	9	1	512

4.3.2 Use of the Binomial

Among litters of four pigs, what proportion will be expected to have all boars, all gilts, 3 boars and 1 gilt, 1 boar and 3 gilts, and so on? First, we will use the binomial expanded to the fourth power (the number of pigs in the litters), and let p represent the probability of a boar and q the probability of a gilt. To calculate the probability that a litter of 4 will have 3 boars and 1 gilt, for example, locate the term containing the factors p^3q (Table 4.2). The coefficient 4 represents the number of different ways (or orders) in which 3 boars and 1 gilt can occur (boar-boar-boar-gilt, boar-boar-gilt-boar, boar-gilt-boar-boar, and gilt-boar-boar-boar). If it is assumed that boars and gilts occur with equal probabilities, we can now calculate the overall probability of obtaining 3 boars and 1 gilt in a litter of 4 pigs:

$$4p^3q = 4(\tfrac{1}{2})^3(\tfrac{1}{2}) = 4(\tfrac{1}{2})^4 = \tfrac{4}{16} = 0.25$$

The usefulness of the binomial is not limited to situations where p and q are each $\tfrac{1}{2}$. Suppose that six calves are produced from matings among cows and bulls all heterozygous for dominant and recessive genes; what is the probability there will be 4 dominant and 2 recessive traits among the six calves? For purposes of this illustration, assume that polledness is the dominant trait and horns the recessive trait. In this example we will use the binomial expanded to the sixth power. Letting p be the probability of polled ($\tfrac{3}{4}$) and the q the probability of horns ($\tfrac{1}{4}$), the term $15\,p^4q^2$ should be used to calculate the overall probability of 4 polled calves and 2 horned calves in a group of 6:

$$15p^4q^2 = 15\,(\tfrac{3}{4})^4(\tfrac{1}{4})^2 = \frac{15\,(81)}{(128)(16)} = \frac{1215}{2048} = 0.593$$

4.4 DETERMINING GOODNESS OF FIT

4.4.1 The Chi-Square Test

Much has been said about the theoretical phenotypic ratios expected from various crosses involving one and two pairs of genes (1 : 2 : 1, 3 : 1, 1 : 1, 9 : 3 : 3 : 1, 1 : 1 : 1 : 1, etc). Every geneticist knows that the results obtained from actual crosses do not always fit the theoretical expected ratios exactly. The question arises as to how much the experimental or observed results can differ from the hypothetical or expected results and still be considered reasonably close to those expectations. If a coin is tossed 100 times, it would not be surprising if the results were not exactly 50 heads and 50 tails. If the results were 45 heads and 55 tails, for example, intuition would suggest that these results are probably reasonable. If 40 progeny were produced from a monohybrid cross where the expected phenotypic ratio was 3 : 1, would we expect exactly 30 dominant phenotypes and 10 recessive phenotypes? How much deviation from the hypothetical ratio would we be willing to accept and still consider that those deviations were merely due to chance? Could results of 32 : 8 considered reasonably close enough to be called a theoretical 3 : 1 ratio? What about 35 : 5, or 28 : 12, or 25 : 15? At some point, our intuitions would not be sufficiently reliable, and intuitions of different observers might cause them to draw different conclusions. To avoid some of these problems, statistical tests are used to help decide how well certain observed results agree with the expected or predicted results. One such test is the chi-square test (for the Greek letter χ).

The chi-square test helps to remove much of the guesswork from the process of deciding whether observed deviations are due merely to chance or to some other cause. It is essentially a mechanism of converting deviations from a hypothetical ratio to a single value called the chi-square value (χ^2). The size of the deviations and the size of the sample are considered in the calculation of the chi-square value. The general formula is

$$\chi^2 = \Sigma \frac{(o - e)^2}{e}$$

where o is the observed value for each class or group in the ratio, e is the expected or predicted value, and the Greek letter Σ means to add all values of

$$\frac{(o - e)^2}{e}.$$

Suppose that 40 progeny were produced from a monohybrid cross, where the expected phenotypic ratio was 3 : 1. Assume that the results actually obtained were 25 dominant phenotypes and 15 recessive phenotypes.

Let us now calculate the χ^2 for these results. The procedure is summarized in Table 4.3.

Suppose that the same deviations had been observed but with 100 progeny instead of 40. Since the squared deviations are expressed as a ratio of the expected values, the chi-square value will be smaller. Deviations of 5 are therefore more significant in a sample of 40 than in a sample of 100. The examples in Table 4.3 involve only two phenotypic classes. Table 4.4 illustrates chi-square calculations for a dihybrid ratio involving four classes.

4.4.2 Use of the Chi-Square Table

Once the chi-square value has been calculated, the next step is to determine the probability that it would be as large as it is due to chance. For this, we will make use of the table of chi-square values (Table 4.5). The table has three parts. Across the top of the table are probability levels; the left side of the table lists degrees of freedom; and the body of the table contains various chi-square values. Degrees of freedom for genetic experiments of this kind are usually one less than the number of classes in the experiment. For the problems listed in Table 4.3, there are two classes; therefore, there is only one degree of freedom. For the problem in Table 4.4, there are three degrees of freedom since four classes exist.

The chi-square analysis of the data in the sample of 100 individuals in Table 4.3 produced a chi-square value of 1.33. Referring to Table 4.5 for one degree of freedom shows that the probability of a chance occurrence lies between 0.30 and 0.20, probably very near 0.25. If this experiment were to be conducted a number of times, deviations of this magnitude could be expected about one-fourth of the time due to chance. Such deviations for a sample of this size are relatively common.

To determine the probability of obtaining a chi-square value as large

TABLE 4.3   Calculations of Chi-Square Values for a 3 : 1
Monohybrid Ratio

	Sample of 40 Individuals		Sample of 100 Individuals	
	Dominant Phenotype	Recessive Phenotype	Dominant Phenotype	Recessive Phenotype
Observed, o	25	15	70	30
Expected, e	30	10	75	25
Deviation, $o - e$	-5	$+5$	-5	$+5$
$(o - e)^2$	25	25	25	25
$(o - e)^2/e$	0.83	2.5	0.33	1.0
χ^2	3.33		1.33	

TABLE 4.4 Calculations of Chi-Square Values for a Dihybrid Ratio of 9 : 3 : 3 : 1[a]

Classes	Observed, o	Expected, e	Deviation, o − e	$(o − e)^2$	$\dfrac{(o − e)^2}{e}$
Dominant-dominant	107	9 × 13 = 117	−10	100	0.85
Dominant-recessive	48	3 × 13 = 39	+9	81	2.07
Recessive-dominant	47	3 × 13 = 39	+8	64	1.64
Recessive-recessive	6	1 × 13 = 13	−7	49	3.77
	208	16 × 13 = 208		χ^2	8.34

[a]The hypothesis is that the ratio 107 : 48 : 47 : 6 does not deviate significantly from a theoretical 9 : 3 : 3 : 1.

TABLE 4.5   Table of Chi-Square

Degrees of Freedom	Probability of a Chance Occurrence						
	0.70	0.50	0.30	0.20	0.10	0.05	0.01
1	0.15	0.46	1.07	1.64	2.71	3.84	6.64
2	0.71	1.39	2.41	3.22	4.60	5.99	9.21
3	1.42	2.37	3.67	4.64	6.25	7.82	11.35
4	2.20	3.36	4.88	5.99	7.78	9.49	13.28
5	3.00	4.35	6.06	7.29	9.24	11.07	15.09
6	3.83	5.35	7.23	8.56	10.64	12.59	16.81
7	4.67	6.35	8.38	9.80	12.02	14.07	18.48
8	5.53	7.34	9.52	11.03	13.36	15.51	20.09

as 8.34 or larger for the sample of 208 individuals in the example in Table 4.4, locate the chi-square value in the third row (three degrees of freedom) of Table 4.5. The exact chi-square value is not in the table, but it would be found somewhere between the values 7.82 and 11.35. The probability of a chance occurrence lies between 0.05 and 0.01. The hypothesis in Table 4.4 stated that the observed ratio was not significantly different than a theoretical 9 : 3 : 3 : 1 ratio. Since the probability that deviations of the magnitude observed would occur due to chance is less than 0.05, usual statistical procedures suggest that the hypothesis should probably not be accepted.

The exact level of significance at which to accept or reject a hypothesis should be determined based on factors and characteristics related to the experiment and best known to the person conducting the experiment. If the stated hypothesis cannot be accepted, the experiment should be reexamined for errors, another appropriate hypothesis might be tested, or some other logical cause should be sought for the significant deviations.

4.4.3 Limitations of the Chi-Square Test

There are at least two important limitations of the chi-square test that should be pointed out: (1) it must be used on the actual numerical data and not on percentages or otherwise modified data; and (2) it cannot be used reliably in an experiment where the expected number within any phenotypic class is less than 5.

STUDY QUESTIONS AND EXERCISES

4.1. Define each term.

binomial	independent event	Pascal's triangle
chi-square	mutually exclusive event	probability
degrees of freedom		

4.2. List some of the ways that a knowledge of probabilities is important to the study of genetics.

4.3. Describe the rule that pertains to the combining of two or more independent events.

4.4. Describe the rule that pertains to the combining of two or more mutually exclusive events.

4.5. Explain how the testcross can be used to test for the presence of a recessive gene in an individual expressing a dominant phenotype, but whose actual genotype is unknown.

4.6. Why is a testcross not always an appropriate method for testing for the presence of a recessive gene?

4.7. Explain why approximately twice the number of progeny are required for testing an individual for the presence of a recessive gene when mating to known heterozygotes compared to mating to homozygous recessives, assuming that the same levels of confidence are desired for each.

4.8. Explain why approximately twice as many progeny are required for testing for the presence of a recessive gene when mating to daughters of known heterozygotes compared to mating to the known heterozygotes themselves, assuming that the same levels of confidence are desired for each.

4.9. Explain why the probability of producing one individual expressing the recessive trait when mating a heterozygote to a group of daughters of known heterozygotes is $\frac{1}{8}$, assuming that one offspring is produced.

4.10. How many terms are in a specific expanded binomial?

4.11. Describe the pattern displayed by the exponents of the p and q factors of the expanded binomial.

4.12. Describe the pattern displayed by the coefficients of the terms in an expanded binomial.

4.13. Describe two methods for determining the coefficients for the terms in any expanded binomial.

4.14. What is meant by the phrase "goodness of fit"?

4.15. How is the chi-square test useful?

4.16. How does sample size affect the size of the chi-square value?

4.17. Explain two ways in which the chi-square test is limited.

PROBLEMS

Probabilities of Nongenetic Events

4.1. What is the probability of obtaining three heads from the toss of three coins?

4.2. If you were to toss three coins simultaneously, say a penny, a nickel, and a dime, how many different ways could you obtain two tails and one head?

4.3. What is the probability that the toss of three coins will produce two heads and one tail?

4.4 What is the probability of obtaining a six on the roll of one balanced die?

4.5. What is the probability of obtaining a two on the roll of two balanced dice?

4.6. How many ways can a seven be obtained on the roll of two balanced dice?

4.7. What is the probability of obtaining a seven on the roll of two balanced dice?

4.8. What is the probability of drawing the ace of spades from a shuffled deck of playing cards? (Excluding jokers, a deck has 52 cards.)

4.9. What is the probability of drawing any ace from a shuffled deck of playing cards consisting of four suits of 13 cards each?

4.10. What is the probability of drawing a king and a queen from a shuffled deck of cards? (Assume that the first card is not replaced in the deck before the second is drawn.)

Probabilities of Genetic Events

4.11. What is the probability that a gamete produced from an individual of the genotype *AaBbcc* will contain the genes *abc*?

4.12. What is the probability that an individual produced from the cross *AaBb* × *Aabb* will possess the genotype *aabb*?

4.13. Polledness (*P_*) and black coat color (*B_*) in cattle are dominant to horns (*pp*) and red coat color (*bb*), respectively. What is the probability that a calf produced from the mating of a dihybrid bull and cow will be **(a)** horned and red; **(b)** polled and red; **(c)** polled and black?

4.14. What is the probability that the dihybrid cross in Problem 4.13 will produce two polled black calves, assuming that two calves are indeed produced?

4.15. Assume that a polled bull is mated to horned cows and produces eight polled calves; if the breeder assumes that the bull is homozygous (*PP*), what is the probability that the assumption is wrong?

4.16. Assume that another polled bull produces five polled calves from matings to known heterozygous cows and six polled calves from daughters of known heterozygotes; what is the probability that the bull is indeed heterozygous and that this would happen due to chance?

4.17. How many polled calves would have to be produced from the mating of a polled bull to horned cows without producing any horned calves to be confident at the 0.999 level that the bull is homozygous?

Probabilities Involving the Binomial Expression

4.18. In litters of five kittens, with what frequency would we expect to find **(a)** 4 males and 1 female; **(b)** 3 males and 2 females; **(c)** 2 males and 3 females; **(d)** 1 male and 4 females; **(e)** all males?

4.19. In litters of seven pigs produced from a sow and boar both of which are heterozygous for dominant black, what is the probability that there will be **(a)** all black pigs; **(b)** all red pigs; **(c)** 2 black pigs and 5 red ones; **(d)** 4 black pigs and 3 red ones?

Chi-Square Tests

4.20. How many degrees of freedom would be associated with testing the following ratios: **(a)** $1 : 2 : 1$; **(b)** $3 : 1$; **(c)** $1 : 1 : 1 : 1$; **(d)** $9 : 3 : 3 : 1$; **(e)** $6 : 4 : 3 : 2 : 1$?

4.21. Dominant and recessive phenotypes are produced in a ratio of $34 : 14$; which theoretical ratio does this fit better, a $3 : 1$ or a $2 : 1$?

4.22. Perform a chi-square analysis to determine how well an observed ratio of $11 : 9$ fits a theoretical ratio of $1 : 1$.

4.23. Would a sample with the same proportions as those reported in Problem 4.22 fit a theoretical ratio of $1 : 1$ if it were **(a)** 10 times larger; **(b)** 20 times larger?

4.24. Within a class of 45 students, each was asked to toss a group of 5 coins five times and record the results; a summary of those results is shown below. How well do these results conform to what would normally be expected?

Proportion of Heads and Tails	Number of Combinations
5H	10
4H, 1T	36
3H, 2T	65
2H, 3T	63
1H, 4T	39
5T	11
	225

REFERENCES

BURNS, G. W. 1980. *The Science of Genetics,* 4th ed. Macmillan Publishing Co., Inc., New York. Chapter 5.

SPIEGEL, M. R. 1975. *Probability and Statistics.* McGraw-Hill Book Company, New York.

STANSFIELD, W. D. 1983. *Theory and Problems of Genetics*, 2nd ed. McGraw-Hill Book Company, New York. Chapters 2 and 7.

VAN VLECK, L. D., E. J. POLLAK, and E. A. B. OLTENACU. 1987. *Genetics for the Animal Sciences*. W. H. Freeman and Company, Publishers, New York. Chapters 5 and 8.

FIVE

Inheritance Related to Sex

Birds and mammals develop from a single diploid cell that originates from the union of two haploid cells. These two haploid cells are designated as male and female gametes and are produced by male and female parents, respectively. The reproductive systems that produce the male and female gametes are limited to one per mating type (sex) in birds and mammals. In other words, male birds and mammals have male reproductive organs that produce only spermatazoa, while females have female reproductive organs that produce only female gametes.

In addition to birds and mammals, many other classes and families of animal life as well as some cultivated plants exhibit only one sex per individual. These kinds of organisms are referred to as *dioecious*. Individuals within some lower forms of animals, such as earthworms, flatworms, mollusks, and echinoderms, and most flowering plants possess complete functional male and female reproductive systems. These are examples of monoecious organisms. The term *hermaphrodite* is sometimes used to describe monoecious organisms, especially among animals.

5.1 SEX CHROMOSOMES AND SEX DETERMINATION

5.1.1 The XX–XY System in Mammals

Of the several pairs of chromosomes in the body cells of mammals, one pair is involved in the process of determining whether the individual is male or female. These are the sex chromosomes that in mammals have been arbitrarily labeled X and Y. Females carry two X chromosomes, while males carry one of each in every diploid body cell.

During spermatogenesis, males produce two kinds of gametes with regard to their sex chromosomes; one kind of gamete carries an X chromosome, the other carries a Y chromosome. Because two kinds of gametes can be produced, the male mammal is known as *heterogametic*. Although there may be factors that cause the ratio of X- to Y-bearing gametes to be unequal, the major factor affecting the ratio is chance. For all practical purposes, therefore, we can assume that on the average the numbers of the two kinds of gametes produced by mammals is approximately equal. Since females carry only X chromosomes, all of their gametes carry an X chromosome. For this reason they are known as the *homogametic* sex.

At fertilization, if the female ovum is fertilized by an X-bearing spermatazoon, a female XX zygote will be produced. If the ovum is fertilized by a Y-bearing sperm cell, a male XY zygote will be produced. The major factor that determines which male gamete will fertilize the ovum is chance. Under most circumstances it can be predicted that approximately one-half of all mammalian zygotes will be genotypically male and one-half will be female. At this level of development, sex is referred to as genetic or *genotypic* sex. Later, when the reproductive tract and the gonads develop and can be distinguished male from female, sex is referred to as *phenotypic*. Many of the genes that control maleness and femaleness are carried on the sex chromosomes and ultimately control the expression of phenotypic sex.

Because of the role that the X and Y chromosomes play in sex determination, some interesting observations can be made regarding them. Male mammals always receive a Y chromosome from their sires and pass on a Y chromosome to their sons. Females always receive an X chromosome from their sires and pass on an X to their sons. Males always receive their X chromosome from their dams.

On rare occasions, gametes can be formed having two X chromosomes, or both an X and a Y chromosome caused by nondisjunction of chromosome pairs during meiosis. Gametes can also be formed that do not contain either sex chromosome. Various zygote types can thus be formed, such as XXY, XYY, XO, YO, XXX, and others. One of these combinations, XXY, produces a condition in humans known as Klinefelter's syndrome. Such individuals are generally of male phenotype but possess some feminine characteristics. They are almost always mentally and reproductively retarded. Another of the types, XO, where the Y chromosome is missing, produces the condition in humans called Turner's syndrome. These individuals are generally of female phenotype, exhibit some mental retardation, and are almost always sterile. From these examples it would seem that in humans, and possibly in mammals in general, phenotypic maleness is probably determined by the presence of the Y chromosome, whereas femaleness is determined by the absence of the Y chromosome.

5.1.2 Sex Determination in Other Animals

The method of sex determination in birds is similar to that in mammals except that the homogametic sex is the male and is usually designated as ZZ instead of XX. The female is the heterogametic sex and is designated as ZW. Male birds obtain a Z chromosome from both parents and pass on a Z chromosome to their daughters. Females obtain a Z chromosome from their sires and pass on a Z to their sons. Female birds always receive a W chromosome from their dams and always pass it on to their daughters. The system of sex determination in some insects and some fishes appears to be similar to that of birds.

The system of sex determination in some insects, including flies, is essentially the same as it is in mammals, with XX females and XY males. In other insects, such as grasshoppers and bugs, females are homogametic (XX) as in mammals. Males are heterogametic, but carry only one sex chromosome, an X. There is no Y chromosome. Males are thus designated X0; the 0 (zero) represents the missing chromosome. One-half the male gametes carry the regular haploid number of chromosomes, including an X. The other one-half of the male gametes carry the haploid number minus one— no sex chromosome is present. In the XX–X0 species, sex seems to be controlled by the number of X chromosomes present; one X produces males, two X's produce females. In mammals sex seems to be controlled by the presence or absence of the Y chromosome. Note that X0 in grasshoppers is a male, but in humans it is a female (Turner's syndrome).

In bees and other members of the order Hymenoptera, males contain the haploid number of chromosomes, while females possess the diploid number. The queen usually mates only once and stores semen in a special pouch in the reproductive tract. Fertilization of the eggs is instinctively controlled by the female at the time of oviposition. Zygotes produced from fertilized eggs are diploid and develop into female adults, the queens and worker bees. The unfertilized zygotes remain haploid and develop into males, or drones.

5.2 SEX-LINKED INHERITANCE

It has already been pointed out that the sex chromosomes carry genes that control masculine and feminine characteristics. However, many traits not otherwise related to sex are also controlled by genes carried on the sex chromosomes.

5.2.1 The Nature of Sex-Linked Genes

Genes carried on the sex chromosomes are said to be *sex linked*, and in the vast majority of cases, this will refer to genes on the X chromosome. In most mammalian species, only a very few traits are known to be controlled by genes carried on the Y chromosome. For that reason, hereafter in this book, unless otherwise indicated, the use of the term *sex linkage* will refer to genes carried on the X chromosome. Genotypes for sex-linked genes in female mammals can be designated in the same way as genotypes for genes on the autosomes. In males only one X chromosome is present, so in general only one gene is present to affect a particular trait. The exception to this is that there is a small region on the X chromosome that is homologous to a small part of the Y chromosome. The homologous regions on the X and Y chromosomes allow for synapsis to occur during meiosis. Loci for any genes that might reside in this homologous region would be expected to exist in pairs and the genes would be expected to segregate as autosomal genes.

The description and functioning of genes carried on the Z and W chromosomes of birds would be very much the same as the previous description regarding the X and Y chromosomes in mammals. Some examples of sex-linked traits in birds and mammals are presented in Table 5.1. Descriptions of a few examples are presented in the following paragraphs.

5.2.2 Colorblindness in Humans

To better understand the effects of sex chromosomes on the genes they carry, let us examine a trait controlled by a pair of X-linked genes. A type

TABLE 5.1   Sex-Linked Traits in Birds and Mammals

Species	Trait	Kind of Inheritance
Human	Colorblindness	Recessive
	Hemophilia A	Recessive
Dog	Hemophilia	Recessive
Cat	Tortoiseshell pattern	Codominance between alleles for yellow and nonyellow
Chicken	Barred feathers	Dominant
	Gold or silver plumage	Dominance for silver
	Rapid feathering in chicks	Recessive
Canary	Green or cinnamon feathers	Dominance for green
	Black or red eyes	Dominance for black
Carrier pigeon	Creamy head	Codominance between alleles for gray and lethality

of colorblindness in humans wherein a person has difficulty distinguishing between red and green is such an example. The gene that controls the colorblind condition is recessive; the dominant allele results in normal vision. Since females carry two X chromosomes, two loci are present to carry the genes, thus two genes control phenotypes in females. If we let an uppercase letter *A* represent the dominant gene for normal vision and a lowercase *a* represent the recessive gene, three phenotypes are possible in female: *AA*, *Aa*, and *aa*. Two of the genotypes result in normal vision (*AA* and *Aa*), while the homozygous recessive *aa* results in colorblindness.

Because males have only one X chromosome, their genotypes would be represented by only one gene each, either *A* or *a*. However, since it is customary to think of genotypes in terms of two symbols, those of the male might be expressed as *AY* and *aY*. The presence of the Y would suggest that the genes *A* and *a* are X linked. Another possibility would be to use the symbols *A0* and *a0*, the zeros indicating the absence of a second gene and again suggesting that the genes are sex linked. Another system that has been used is to designate the X and Y chromosomes and note the gene symbols as subscripts on the X's: $X_A X_a$, $X_a X_a$, $X_A Y$, and so on.

For a woman to express colorblindness, she must carry a recessive gene on both X chromosomes. The situation is the same as if the genes were on a pair of autosomes. In men, since there is only one X chromosome, colorblindness will be expressed when only one recessive gene is present. For a woman to be colorblind, her father must also have been colorblind, and her mother must at least be heterozygous, as in the following illustration:

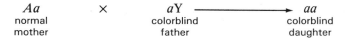

Aa	×	*a*Y	⟶	*aa*
normal		colorblind		colorblind
mother		father		daughter

We can state further that if the colorblind daughter were to have children, all of her sons would receive a recessive gene and thus be colorblind. And assuming that the daughter's husband were not colorblind, all their daughters would have normal vision but would be carriers.

aa	×	*A*Y	⟶	*a*Y	and	*Aa*
colorblind		normal		colorblind		carrier
woman		man		sons		daughters

5.2.3 Hemophilia in Dogs

Hemophilia is a genetic disease caused by a deficiency of one of the many factors that controls the blood-clotting reaction. In both dogs and humans, this particular type of hemophilia is known as hemophilia A. The deficient substance has been identified by physiologists as Factor VIII. The production of this factor is under control of a dominant gene carried on the X chromosome. Females that are homozygous recessive and males carrying the recessive gene (sometimes referred to as *hemizygous*) on their X chromosome express hemophilia A since the dominant gene that controls the production of Factor VIII is absent.

Dogs that express the recessive trait seldom live long enough to reproduce. This means that matings that could produce hemophiliac pups would be limited to normal males (AY) bred to carrier females (Aa). One-half the male progeny from this cross would be expected to be aY and express hemophilia. The other one-half would be expected to be AY and normal. One-half the females would be expected to be AA and one-half Aa. All females should be pheontypically normal.

5.2.4 The Tortoiseshell Cat

The examples of sex-linked inheritance presented thus far have involved alleles exhibiting complete dominance and recessiveness. An interesting situation exists with X-linked alleles that are codominant. Since the heterozygote cannot exist in males, females can express a phenotype that males cannot. One such example involves genes that result in the tortoiseshell pattern in cats.

Orange (or yellow) is caused by the homozygous state of a gene O in female cats that would otherwise have been black. The expression of black is possible due to the presence of the allele o in the homozygous state (oo). Males may carry either one gene for orange (OY), or one gene that allows black to be expressed (oY). Heterozygous female cats, Oo, express the so-called tortoiseshell pattern. This pattern is characterized by the presence of both black and yellow hairs, producing a blotchy appearance similar to the roan pattern in cattle. Because males do not normally carry two X chromosomes, they would not be expected to be heterozygous and therefore the tortoiseshell pattern would not be expected in male cats. All possible crosses among cats for these colors are listed in Table 5.2.

Depending on other genes that may be present, cats may also have white spots. When white spots occur with heterozygous tortoiseshell genotypes, the black and orange colors tend to be expressed as separate spots rather than as the blotchy tortoiseshell pattern. The resulting three-color pattern is commonly known as calico.

As was pointed out earlier, the tortoiseshell or calico patterns do not normally occur in male cats. However, a very small number of such phenotypes have been reported. For the tortoiseshell pattern to exist, both a gene for orange and a gene for black must be present in the individual. This would mean that two X chromosomes must be present. For the cat to be a

TABLE 5.2   **Phenotypes and Crosses Involving Black, Orange, and Tortoiseshell Colors in Cats**

Male Parent	Female Parent	Progeny
Black	Black	All black males and females
	Orange	Orange males, tortoiseshell females
	Tortoiseshell	Males: $\frac{1}{2}$ black, $\frac{1}{2}$ orange
		Females: $\frac{1}{2}$ black, $\frac{1}{2}$ tortoiseshell
Orange	Black	Black males, tortoiseshell females
	Orange	All orange males and females
	Tortoiseshell	Males: $\frac{1}{2}$ orange, $\frac{1}{2}$ black
		Females: $\frac{1}{2}$ orange, $\frac{1}{2}$ tortoiseshell

male, a Y chromosome must be present. So one explanation of the existence of a tortoiseshell or calico cat is that it would carry an extra X chromosome. Recall that this XXY condition is the one that resulted in Klinefelter's syndrome in humans. Such an individual expresses the male phenotype and is usually reproductively sterile. Male cats with this aberration are almost always sterile.

5.3 SEX-INFLUENCED INHERITANCE

A number of traits are known in humans and domestic animals wherein the nature of the interaction between alleles depends on the sex of the individual. The genes are carried on the autosomes rather than on the X chromosomes. Genotypes would therefore be the same in males and females. Interaction between alleles is such that one is usually dominant in males and recessive in females. The allele that is recessive in males is usually dominant in females. The manner in which the alleles interact is influenced by the presence of sex hormones in the respective sexes. Again, the genes are not carried on the X chromosomes. After having learned about *sex-linked inheritance*, many students also have a tendency to associate the genes for sex-influenced traits with the X chromosomes.

    The mahogany and red colors in cattle represent a good example to illustrate sex-influenced inheritance. Mahogany is a very deep red, almost maroon. It is often seen in Maine-Anjou and Ayrshire cattle. Red in cattle is that color usually seen in Herefords and Shorthorns. White spotting is almost always associated with red or mahogany colors, but the genes for white spotting are inherited independently of those for red or mahogany. Mahogany color is controlled by a gene completely dominant in bulls and completely recessive in cows and heifers. On the other hand, the usual red color is controlled by an allele dominant in females and recessive in bulls.

    Using the symbols *M* and *m* to represent the genes for mahogany and red, respectively, it can be seen that three different genotypes can exist. The homozygote *MM* would produce mahogany color in both males and females. The homozygous *mm* would result in red color in both males and females. As has been illustrated on a number of occasions previously, whether alleles are dominant, recessive, or codominant has no bearing on the expression of a homozygote. The phenotypic expression of a heterozygote, however, is a different matter. With complete dominance and recessiveness, a heterozygote expresses itself in the same way as the dominant homozygote. With codominance, the phenotype of the heterozygote is different from either of the homozygotes. With sex-influenced traits such as color in Ayrshires, the expression of the heterozygote depends further on the sex of the individual. In this case the heterozygote *Mm* would be mahogany in bulls and red in females. Table 5.3 shows the various genotypes, phenotypes and crosses related to these two colors in cattle.

Other examples of traits that express themselves in a manner similar to mahogany and red in cattle include a type of baldness in humans, scurs in polled cattle, beards in goats, and horns in sheep.

5.4 SEX-LIMITED TRAITS

In a broad sense, a *sex-limited trait* is any trait that is normally expressed in one sex and not in the other. Milk production in mammals and egg production in birds are two good examples. In these examples, the expression

TABLE 5.3   Genotypes, Phenotypes, and Crosses Related to Mahogany and Red Colors in Ayrshire Cattle

Bulls		Cows	Calves
Mahogany	×	Mahogany	
MM	×	*MM*	All calves mahogany
Mm	×	*MM*	All bulls mahogany, $\frac{1}{2}$ heifers mahogany, $\frac{1}{2}$ heifers red
Red	×	Red	
mm	×	*mm*	All calves red
mm	×	*Mm*	All heifers red, $\frac{1}{2}$ bulls mahogany, $\frac{1}{2}$ red
Red	×	Mahogany	
mm	×	*MM*	Mahogany bulls, red heifers
Mahogany	×	Red	
MM	×	*mm*	Mahogany bulls, red heifers
Mm	×	*mm*	All heifers red, $\frac{1}{2}$ bulls mahogany, $\frac{1}{2}$ red
MM	×	*Mm*	All bulls mahogany, $\frac{1}{2}$ heifers mahogany, $\frac{1}{2}$ heifers red
Mm	×	*Mm*	$\frac{3}{4}$ bulls mahogany, $\frac{1}{4}$ red, $\frac{3}{4}$ heifers red, $\frac{1}{4}$ mahogany

of the trait is dependent on the development of certain anatomical structures, which in turn produce the particular product. Many pairs of genes are involved, so the discussion of this kind of trait will be reserved for another chapter.

The cock-feathering trait in birds is a sex-limited trait controlled by only one pair of autosomal genes. A dominant gene *H* produces hen feathering in both hens and cocks. Since the gene is dominant to its allele, hen-feathering genotypes can be homozygous (*HH*) or heterozygous (*Hh*). The recessive genotype (*hh*) may be present in both sexes, but cock feathering is expressed only in males. The presence of the male hormone is apparently necessary for the cock-feathering gene to express itself.

Note the differences in sex-limited inheritance and sex-influenced inheritance. With sex-limited inheritance, the gene dominant in the male is also dominant in the female; this is not the case with sex-influenced inheritance. With the cock-feathering trait, the effect of the sex hormone is to determine whether the trait will or will not be expressed. In sex-influenced inheritance, the effect of the sex hormones is to determine the dominance or recessiveness of the alleles. A similarity is that the genes for both kinds of traits are carried on the autosomes.

Another example of a sex-limited trait involves a series of multiple alleles that affects the development of horns in sheep. This trait is described in Chapter Six.

STUDY QUESTIONS AND EXERCISES

5.1. Define each term.

| dioecious | heterogametic | monoecious |
| hermaphrodite | homogametic | nondisjunction |

5.2. Describe the differences in sex chromosomes of birds and mammals.

5.3. What is the primary factor that determines the sex of a particular individual?

5.4. Provide one possible explanation for how Klinefelter's syndrome can occur.

5.5. Provide one possible explanation for how Turner's syndrome can occur.

5.6. Provide a possible explanation for how a male tortoiseshell cat can be produced.

5.7. Distinguish between sex-linked and sex-influenced traits.

5.8. Distinguish between sex-influenced and sex-limited traits.

PROBLEMS

5.1. Red-green colorblindness in humans is caused by a sex-linked recessive gene; the dominant gene results in normal vision. In addition, assume that brown eye color is controlled by an autosomal dominant gene and blue eyes by the recessive allele when present in the homozygous state. **(a)** A woman with blue eyes and normal vision has a colorblind father; the woman's husband has normal vision and brown eyes but his mother had blue eyes; assume that all their children are boys; what are the possibilities regarding eye color and colorblindness? **(b)** Assume that all the children of the couple in part (a) were girls; what are the possibilities with regard to eye color and colorblindness?

5.2. The tortoiseshell color pattern in cats is the result of an interaction between two sex-linked codominant alleles. One of the alleles controls the production of black, the other controls the production of yellow. White spotting is controlled by an autosomal recessive gene, while the presence of the dominant allele results in the absence of spotting. When white spotting occurs with the tortoiseshell pattern, it is called calico. **(a)** Assume that a calico female was bred to a solid black male that carried a recessive spotting gene. What colors and patterns would be expected among the kittens from this mating, and in what proportions? **(b)** A tortoiseshell female (not calico) has a litter of seven kittens, including 2 solid yellow females, 1 tortoiseshell female, 1 calico female, 1 solid black male, 1 solid yellow male, and 1 spotted yellow male; describe the genotype and phenotype of the sire of this litter.

5.3. In carrier pigeons a sex-linked gene *G* produces the gray-headed trait; a codominant allele *g* is lethal when present in the homozygous state in males. Since females are heterogametic, having only one Z chromosome, the presence of just one *g* allele is a lethal condition. The heterozygous male has a creamy-colored head. **(a)** List all genotypes and phenotypes that can exist in mature pigeons with regard to the *G* locus. **(b)** List the genotypes and phenotypes of all possible progeny expected from a cross between gray-headed females and creamy-headed males. **(c)** List the genotypes and phenotypes of all possible progeny expected from a cross between gray-headed males and gray-headed females. **(d)** Is any other kind of cross possible? If so, indicate the cross and results expected in the progeny.

5.4. A dominant sex-linked gene in chickens produces barred feathers; the recessive allele controls the expression of nonbarred plumage. The creeper condition is the result of the interaction between two codominant autosomal alleles in the heterozygote. One of the alleles is lethal when homozygous, the other results in birds with normal legs. **(a)** If barred creeper females were mated to nonbarred creeper males, what phenotypes would be expected among the progeny, and in what proportions? **(b)** Assume that a normal nonbarred hen produced a brood consisting of 4 barred creeper males, 2 barred normal males, 2 barred creeper females, and 3 barred normal females; describe the most likely genotype and phenotype of the sire.

5.5. The mahogany and red colors in Ayrshire cattle are controlled by a pair of sex-influenced autosomal alleles. The gene that produces mahogany is dominant in males and recessive in females. The allele for red is dominant in females and recessive in males. Polledness in cattle is produced by a dominant gene, while the horned trait is the result of a recessive allele when it exists in the homozygous state. **(a)** An Ayrshire breeder whose entire herd is made up of horned mahogany cows decides to breed all of them to red polled bulls. If the polled red bulls are homozygous for the polled trait, what phenotypes can be expected among the calves from these crosses? **(b)** What are the probable expectations from dihybrid crosses involving the genes at these two loci? **(c)** Describe how a purebred (homozygous) herd of polled mahogany cattle could be produced.

5.6. The gene for cock feathering in chickens is recessive to the dominant allele for hen feathering. The gene for cock feathering is expressed only in males; hen feathering can be expressed in males and females. Silver-colored feathers are due to the expression of a dominant sex-linked gene, while gold plumage is controlled by the recessive allele. **(a)** A gold, cock-feathered male is bred to a silver-colored hen that is known to be heterozygous for the cock- and hen-feathering genes; what is the probability that male progeny will be gold and cock feathered? **(b)** What genotypes and phenotypes would be expected among the female progeny of the cross in part (a)? **(c)** What phenotypic result would be expected from crosses between hens and cocks, both of which have gold plumage and are heterozygous for hen feathering?

REFERENCES

CENTERWALL, W. R., and K. BENIRSCHKE. 1973. Male tortoise-shell and calico cats. *Journal of Heredity* 64 : 272.

HUTT, F. B. 1949. *Genetics of the Fowl*. McGraw-Hill Book Company, New York.

HUTT, F. B., C. G. RICKARD, and R. A. FIELD. 1948. Sex-linked hemophilia in dogs. *Journal of Heredity* 39 : 2.

NOVITSKI, E. 1982. *Human Genetics*, 2nd ed. Macmillian Publishing Co., Inc., New York.

STERN, C. 1973. *Principles of Human Genetics*, 3rd ed. W.H. Freeman and Company, Publishers, San Francisco.

SIX

Multiple Alleles

6.1 NATURE OF MULTIPLE ALLELES

Alleles were defined earlier as the different kinds of genes that can occupy the same loci on homologous chromosomes. An allele can be thought of as an alternate form of another gene, one that was possibly formed by a mutation. Problems and illustrations used in previous chapters involved only two alleles. It should be evident that if one allele can be formed by way of a mutation, any number of such changes could occur, producing any number of alleles. When more than two alleles affect a given trait, that trait is said to be controlled by multiple alleles.

Only two alleles can exist in a single individual since there are only two loci on a pair of homologous chromosomes where those alleles can exist. In a population, any number of alleles can exist. The theoretical number of alleles that could exist in a population is two times the number of individuals in the population. Some traits may be controlled by over 300 alleles; one of the blood group systems in cattle is an example.

6.2 COLOR PATTERNS IN CATTLE

The three most common breeds of cattle exhibit three different color patterns. Hereford cattle show the distinctive white-face pattern that also includes white on the underside of the body and usually on the top of the neck (see Figure 6.1). This pattern is so distinctive among Herefords that it is usually referred to as the Hereford pattern. Angus cattle show no white and are usually referred to as being solid colored. The third pattern is the spotted pattern seen in Holstein cattle. Whether cattle are black or red is not a factor in the expression of the Hereford, solid, or spotted patterns.

When purebred Hereford cattle assumed to be homozygous are mated to purebred Angus cattle, the calves express the white-face pattern. If the F_1 cattle are allowed to mate among themselves, it can be observed that the F calves will express the two parental patterns in the ratio of $\frac{3}{4}$ white face to $\frac{1}{4}$ solid. This is a demonstration that these two patterns are controlled by two alleles, with the one controlling the Hereford pattern being completely dominant over the one for the solid pattern.

When purebred Angus cattle are mated to purebred Holstein cattle, the calves are usually all solid black and heterozygous. Crosses among black cattle of this kind produce calves $\frac{3}{4}$ of which are also solid colored, while the other $\frac{1}{4}$ of the calves show the spotted pattern. These crosses indicate that the solid pattern is dominant to spotting. Similar results are obtained when crosses are performed between Holsteins and Herefords—the F_1 calves usually express the white-face pattern of the Hereford. The F_2 generation will be composed of approximately 75 percent white-faced calves and 25 percent calves with Holstein spotting, indicating that the gene for the Hereford pattern is dominant to the one for white spotting.

From the foregoing discussion, it can be concluded that the three patterns are controlled by three alleles, one (white face) being completely dominant to the other two, and one (white spotting) that is completely recessive to the other two. The symbols S^h, S, and s can be used to represent the three alleles controlling the traits Hereford pattern, solid, and spotted, respectively. The hierarchy of interaction can be expressed this way: $S^h > S > s$, with any gene being dominant to those to the right, and recessive to those to the left. Figure 6.1 shows the three color patterns just described.

There is some evidence that a fourth allele at this locus produces the color-sided pattern common in the Pinzgauer breed of cattle (see Figure 6.2). Colored hair is present on the head, neck, and sides of the body. A white strip extends from the withers down the back and behind the animal, continuing along the underline to the brisket. The allele responsible, symbolized as S^c, appears to be codominant with the gene for Hereford pattern, and completely dominant to the other two alleles. Cattle heterozygous for the genes for Hereford and color-sided patterns should express both patterns. Table 6.1 lists the genotypes and phenotypes associated with the four alleles at the white spotting locus.

Figure 6.1 Color patterns commonly observed in cattle. Top: Hereford pattern (Owner: Curtis Reynolds, Stark City, MO); middle: solid color with no white markings (Owner: John Sparkman, Republic, MO); bottom: recessive white spotting (Owner: Gordon Wilson, Republic, MO).

Figure 6.2 Color-sided pattern in a Pinzgauer bull. (Courtesy of New Breeds Industries, Inc., Manhattan, KS).

6.3 MULTIPLE ALLELES IN HORSES

Genes at several loci control the many colors and patterns in horses. A more detailed description of their inheritance will be found in Chapter Eight. One series of colors is thought to be controlled by at least three alleles. A recessive gene in the homozygous state (*aa*) allows for the expression of black hair over the whole body. A second allele, *a^t*, which is dominant to the gene

TABLE 6.1   **Genotypes and Phenotypes for Several**
Color Patterns in Cattle

Homozygous Genotypes	Heterozygous Genotypes	Phenotypes
S^hS^h	S^hS, S^hs	Hereford pattern
S^cS^c	S^cS, S^cs	Color-sided
—	S^hS^c	Color-sided and white-faced
SS	Ss	Solid color
ss	—	Holstein spotting

for black, results in the expression of seal brown. Seal brown is similar to black except that lighter brown or tan hair appears over the muzzle and eyes, and in the area of the flanks. Two possible genotypes allow for the expression of seal brown, the homozygote $a^t a^t$ and the heterozygote $a^t a$.

The third allele (A) in this series is dominant to each of the other two alleles and results in the bay pattern when present in the genotype. Bay is a pattern where black pigment shows in the hair of the mane, tail, and lower part of the legs. The rest of the body is usually brown or reddish brown in color. Since the so-called bay gene is completely dominant to its two alleles, three genotypes are possible for this pattern: AA, Aa^t, and Aa.

6.4 THE ABO BLOOD GROUP SYSTEM IN HUMANS

One of several blood group systems in humans is called the ABO system. It is controlled by three alleles—A, B, and a. The alleles A and B express codominance between them and both are completely dominant to a. The gene A controls the production of a substance in the blood called antigen A. The gene B controls the production of antigen B. The recessive allele results in the absence of either of these antigens. Table 6.2 lists the genotypes and blood groups possible from these three alleles.

6.4.1 Importance in Blood Transfusions

If blood containing the A or B antigen is introduced into the blood of a person where that particular antigen does not normally exist, the introduced antigen stimulates the production of certain substances in the body called *antibodies*. These antibodies bring about the agglutination of the red blood cells. This clumping process can cause a plugging of the capillaries and small arteries, resulting in serious illness and even death. Because of this antigen–antibody reaction, it is important for people to know to which ABO blood group they belong.

People with group O blood can receive only group O blood because

TABLE 6.2   Genotypes and Blood Groups in the ABO System

Blood Group	Genotypes	Antigens in the Blood	Antibodies in the Blood
O	aa	None	Anti-A, anti-B
A	AA, Aa	A	Anti-B
B	BB, Ba	B	Anti-A
AB	AB	A and B	None

neither antigen is normal in their blood. But since their blood contains no antigens, they can donate blood to someone of any group. For this reason, they are called *universal donors* (see Figure 6.3). People with group A blood cannot receive blood that contains any B antigens (B or AB). And since their blood contains antigen A, they cannot donate blood to anyone with group O or group B blood. They can donate to someone with A or AB blood since the A-antigen is normally present.

Those with group B blood should not accept blood from anyone in groups A or AB because their blood contains A antigens. Neither should they donate to anyone that does not normally have antigen B in their blood: namely, groups A and O. Because the blood of group AB people contains both antigens A and B, their blood should not be given to anyone who is not also in the AB group. And since both types of antigens are normal for their blood, those with AB blood can accept blood from any of the four groups. For this reason, they are called *universal recipients*.

A fortunate situation exists with regard to the relative frequencies of the four groups of the ABO system. Something on the order of 40 percent of the human population in the United States has group O blood. Not being able to receive any other blood but group O, then, does not create a grave disaster. The rarest of the four groups in the population is AB. But since people with this kind of blood can receive blood of any other group, they, too, are in a fortunate position when it comes to blood transfusions.

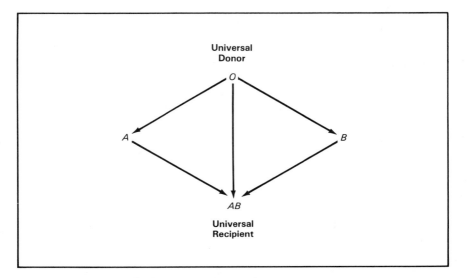

Figure 6.3 Illustration of how the ABO blood groups in humans affect transfusions.

6.4.2 Medical and Legal Applications

Another way in which a knowledge of blood groups can be useful relates to our ability to use such information to determine whether certain individuals may or may not be related. Table 6.3 lists all possible combinations of blood groups in parents and which groups are and are not possible among their children. One assumption that must be made is that parents of either group A or group B must be assumed to be heterozygous; homozygosity for a dominant gene can never be positively guaranteed.

Several interesting facts can be observed from the study of Table 6.3. No person with group O blood can be the parent of a child with group AB blood; conversely, no person with AB blood can be the parent of a child with group O blood, regardless of the blood group of the other parent. Only one combination of blood groups in parents can produce children of all four groups—A × B.

Now consider the situation where a woman has a child and there is a dispute as to who may or may not be the father. A study of Table 6.4 will show that with certain combinations of blood groups for a mother and child, some blood groups are not possible for the father.

6.5 MULTIPLE ALLELES WITH LETHAL EFFECTS

In two of the foregoing examples of multiple alleles, each was related to the others with complete dominance and recessiveness. In one example, both complete dominance and codominance were involved. In the next two ex-

TABLE 6.3   Phenotypes of All Possible Family Combinations
for the ABO Blood Group System

	Children	
	---	---
Parents	Groups That Are Possible	Groups That Are Not Possible
O × O	O	A, B, AB
O × A	O, A	B, AB
O × B	O, B	A, AB
O × AB	A, B	O, AB
A × A	O, A	B, AB
B × B	O, B	A, AB
A × AB	A, B, AB	O
B × AB	A, B, AB	O
AB × AB	A, B, AB	O
A × B	O, A, B, AB	—

TABLE 6.4   Medical–Legal Applications of ABO Blood Group

Blood Group of Mother	Blood Group of Child	Blood Group to Which Father Cannot Belong
O	O	AB
O	A	O, B
O	B	O, A
A	O	AB
A	B	O, A
A	AB	O, A
B	O	AB
B	A	O, B
B	AB	O, B
AB	AB	O

amples we shall see how genotypes for multiple alleles are affected when one or more of the alleles also carry lethal effects.

6.5.1 Coat Color in Foxes

An interesting set of traits in domestic foxes involves three alleles that interact by means of codominance or incomplete dominance. In addition, two of the alleles have lethal effects on the embryos that carry them.

Three symbols (W, w, and p) can be used to represent the three alleles. Three colors are produced: WW results in silver fur, Ww produces white-faced silver, and Wp produces the popular platinum color. Notice that the white-faced silver and platinum colors can only result from heterozygous genotypes. Table 6.5 lists the six possible crosses among foxes of the three colors and the expected theoretical ratios among their progeny.

The genes w and p produce lethal effects both in the homozygous states, ww and pp, and as the heterozygote wp. Each of the genotypes usu-

TABLE 6.5   Crosses among Three Colors of Domestic Foxes

Genotypes of Parents	Genotypes of Zygotes	Phenotypes of Living Progeny at Birth
$WW \times WW$	All WW	All silver
$WW \times Ww$	$\frac{1}{2}WW : \frac{1}{2}Ww$	$\frac{1}{2}$ silver : $\frac{1}{2}$ white-faced silver
$WW \times Wp$	$\frac{1}{2}WW : \frac{1}{2}Wp$	$\frac{1}{2}$ silver : $\frac{1}{2}$ platinum
$Ww \times Ww$	$\frac{1}{4}WW : \frac{1}{2}Ww : \frac{1}{4}ww$	$\frac{1}{3}$ silver : $\frac{2}{3}$ white-faced silver
$Wp \times Wp$	$\frac{1}{4}WW : \frac{1}{2}Wp : \frac{1}{4}pp$	$\frac{1}{3}$ silver : $\frac{2}{3}$ platinum
$Wp \times Ww$	$\frac{1}{4}WW : \frac{1}{4}Ww : \frac{1}{4}Wp : \frac{1}{4}Wp$	$\frac{1}{3}$ silver : $\frac{1}{3}$ white-faced silver : $\frac{1}{3}$ platinum

ally results in the death of the embryos during the early stages of pregnancy. On a few occasions when platinum foxes were mated together ($Wp \times Wp$), a pure white pup was produced among the litters, but died within a few hours after birth. The probable explanation is that the p allele in the homozygous state, while normally causing the embryo to die, failed to kill the embryo as early as expected. It also suggests that if this gene were not lethal, it would probably produce foxes with white fur.

6.5.2. The Agouti Locus in Mice

The gene controlling the agouti pattern in mice has at least four other alleles. The five alleles may be represented by the symbols in Table 6.6. In the homozygous state, the gene for yellow is similar to genes for the creeper trait in chickens and the hairless trait in dogs in that it produces a lethal condition in the homozygous state. The yellow trait is produced only from the heterozygous state of the gene for yellow with one of the nonyellow alleles. Matings among yellow mice produce litters composed of approximately two-thirds yellow and one-third nonyellow mice. In addition, the size of the litters from these matings is usually reduced by about 25 percent. The gene for yellow thus does not exhibit complete dominance to the other four alleles. These other alleles apparently exhibit dominance among themselves in the order listed in Table 6.6. Only five alleles have been identified here as belonging to the agouti locus. Some sources list as many as 13.

6.6 THE ALBINO LOCUS IN RABBITS AND MICE

The albino gene exists in most mammals. It is a recessive mutant of the gene that controls the full production of melanin, the source of all color in skin, hair, and eyes. In mice, at least five alleles exist at this locus. They are listed in Table 6.7 generally in order of dominance.

In rabbits, four of the alleles at the C locus are the same as those

TABLE 6.6   Some of the Alleles at the Aquoti Locus in Mice

Allele	Genotypes	Phenotypes
A^y	$A^y A^y$	None (embryonic lethal)
	$A^y\_\_$	Yellow
A^w	$A^w A^w,\ A^w A,\ A^w a^t,$	White-bellied agouti
	$A^w a$	
A	$AA,\ Aa^t,\ Aa$	Solid agouti
a^t	$a^t a^t,\ a^t a$	Black with tan belly
a	aa	Nonagouti (black)

TABLE 6.7   Alleles, Genotypes, and Phenotypes Associated with the Albino Locus in Mice

Alleles	Genotypes	Phenotypes
C	CC, Cc^w, Cc^h, Cc^e, Cc	Full color
c^c	$c^c c^c, c^c c^h, c^c c^e, c^c c$	Chinchilla
c^h	$c^h c^h, c^h c^e, c^h c$	Himalayan
c^e	$c^e c^e, c^e c$	Extreme dilution
c	cc	Albino

listed for mice, all but c^e. In addition, there is evidence to indicate that two additional alleles affect the degree of expression of the chinchilla pattern. The six alleles listed in order of dominance from left to right are $C > c^d > c^c > c^l > c^h > c$. Table 6.8 lists these alleles and the genotypes and phenotype associated with them.

6.7 MULTIPLE ALLELES RELATED TO SEX

6.7.1 Sex-Linked Multiple Alleles in Pigeons

Three alleles located at one locus on the Z chromosome in pigeons control three feather colors ash red, blue, and chocolate. For simplicity, we shall represent the genes for these colors with the symbols A, B, and b, respectively. The gene for ash red is completely dominant to the other two alleles, while the allele for blue is dominant to the one for chocolate. The genotypes and phenotypes are shown in Table 6.9.

6.7.2 Horns in Sheep

When purebred (homozygous) horned Dorset ewes are bred to horned Merino or Rambouillet rams, the F_1 progeny eventually develop horns in both sexes. The F_1 hybrids mated among themselves constitute monohybrid

TABLE 6.8   Genotypes and Phenotypes Associated with Genes at the Albino Locus in Rabbits

Alleles	Genotypes	Phenotypes
C	$CC, Cc^d, Cc^c, Cc^l, Cc^h, Cc$	Full color
c^d	$c^d c^d, c^d c^c, c^d c^l, c^d c^h, c^d c$	Dark chinchilla
c^c	$c^c c^c, c^c c^l, c^c c^h, c^c c$	Chinchilla
c^l	$c^l c^l, c^l c^h, c^l c$	Light chinchilla
c^h	$c^h c^h, c^h c$	Himalayan
c	cc	Albino

TABLE 6.9   Genotypes and Phenotypes for Feather Colors in Pigeons

Genotypes		
Males	Females	Phenotypes
AA, AB, Aa	AW	Ash red
BB, Ba	BW	Blue
aa	aW	Chocolate

crosses and would be expected to produce theoretical genotypic ratios of
$1 : 2 : 1$. The phenotypes among the female F_2 progeny develop at the rate
of $\frac{3}{4}$ horned : $\frac{1}{4}$ polled. These results suggest that the character is controlled
by two alleles, one for the horned trait that is completely dominant to the
one for the polled trait. However, among the F_2 male progeny, all develop
horns. This further suggests that the expression of the polled trait is limited
to the female and that males are always horned regardless of their genotype.

When Merino or Rambouillet rams (horned) are mated to purebred
polled Suffolks, Shropshires, or Southdowns, the F_1 progeny usually always
remain polled as they grow older. Crosses among the F_1 hybrids produce
females all of which are polled, and males one-fourth of which develop
horns. These results suggest two alleles, the one for horns being completely
recessive to the other and whose expression is limited to the male.

In summarizing the two foregoing examples, there is the suggestion
that three alleles are possibly in control of the horned and polled traits in
sheep. Most Dorsets appear to be homozygous for a dominant gene for
horns that is expressed in both sexes, whereas most other medium-wool
breeds, such as Southdowns, Shropshires, Suffolks, and probably Hampshires, carry a second dominant gene for polledness that is expressed in both
ewes and rams. A third gene that is apparently an allele of the other two
genes seems to be completely recessive to both. This recessive allele is carried in the fine-wool breeds, such as Merino and Rambouillet, in the homozygous state. However, the expression of the gene is affected by the sex of
the individual, such that rams are always horned and ewes are always
polled. The two kinds of crosses described in the foregoing examples illustrate sex-limited inheritance similar to that of the hen- and cock-feathering
traits in chickens described in Chapter Five.

If the symbol h is used to represent the recessive gene in the fine-wool
breeds, their genotype would be hh. The genotypes for horned Dorsets
could be represented by the symbol H in the homozygous state (HH). However, it is possible for Dorsets to be heterozygous and be horned. The other
medium-wool breeds could be either homozygous or heterozygous for the
dominant gene for polled represented by the symbol P, the genotypes being
PP or Ph.

The question then arises as to the nature of the interaction between the dominant gene for horns (*h*) in Dorsets and the dominant gene for polledness (*P*) in the other medium-wool breeds. Crosses among members of these breeds and among their F₁ progeny should answer the question. Matings between homozygotes of the Dorset breed and homozygotes from any one of the other breeds, such as Suffolks, Shropshires, or Southdowns, produce ram lambs that develop horns and ewe lambs that usually never develop horns. Matings among these F₁ hybrids produce a classic example of sex-influenced inheritance similar to the example of the mahogany and red colors in Ayrshire cattle described in Chapter Five. The two alleles *H* and *P*, then, are influenced by sex, such that the polled gene *P* is completely dominant in ewes, but recessive in rams. The horned gene is conversely dominant in rams and recessive in ewes. A further interesting note is that the horned gene in the heterozygous state in rams does not produce horns as large as those of the homozygotes. This situation could therefore be interpreted as a case of partial dominance for the horned gene in the male. The presence of the polled gene appears to in some way affect the development of the horns. The several genotypes and phenotypes controlled by the three alleles are shown in Table 6.10.

PROBLEMS

6.1 The phenotypes controlled by the genes at the agouti locus in mice were explained previously in this chapter. Suppose that yellow mice of the genotype *AʸA* were crossed with agouti mice of the genotype *Aa*: **(a)** List the genotypes and theoretical ratios that should result. **(b)** What ratio of phenotypes would be expected? **(c)** Predict and explain the phenotypic results of a cross between the two kinds of yellow mice from part (a).

6.2. At least three color patterns may exist among Mallard ducks and are controlled by three alleles: mallard (*M*), restricted mallard (*m*), and dusky (*d*). When homozygous mallards are mated to ducks homozygous for the dusky pattern, further crosses among the F₁ produce an F₂ generation with three-fourths of them expressing the mallard pattern and one-fourth expressing the dusky pat-

TABLE 6.10   Genotypes and Phenotypes for Horns and Polledness in Sheep

Genotypes	Phenotypes
PP or *Ph*	Polled in both sexes as in most medium-wool breeds except Dorsets
HH	Horns in both sexes—Dorsets
HP	Polled ewes, small horns in rams
Ph	Horns in both sexes
hh	Horned rams, polled ewes—Merinos and Rambouillets

tern. When ducks with the restricted mallard pattern are mated to ones with the dusky pattern, crosses among the F_1 produce an F_2 generation with three-fourths of them expressing the restricted mallard pattern and one-fourth expressing the dusky pattern. **(a)** If homozygous ducks with the restricted mallard trait were mated to homozygotes expressing the mallard pattern, what would you expect from crosses among the F_1 generation if only two phenotypes could be expressed? **(b)** Assume that the gene for mallard is dominant to the allele for restricted mallard; calculate the theoretical results of mating a drake of the genotype *Mm* to a group of homozygous females of equal numbers of mallards, restricted mallards, and duskies.

6.3. The genetics of the ABO blood group system in humans was explained previously in this chapter. **(a)** Assume that a man with group A blood has been brought into a paternity suit by a woman who also has group A blood; her child has AB blood. What is your conclusion as to whether or not this man is the father of the child? **(b)** Suppose that three babies were born on the same night in a small hospital. Their blood groups are O, A, and AB, and the three sets of parents are AB × O, A × O, and B × AB. Assign the three babies to their respective parents using this information.

6.4. The MN blood group system in humans is controlled by two codominant alleles, *M* and *N*. The genotypes *MM*, *MN*, and *NN* produce the three blood groups M, MN, and N, respectively. **(a)** What is the probability that a baby's blood type will be AN if the parent are AB MN and ON? **(b)** A man whose blood type is A MN and his wife whose type is AN have a child with group O blood. If this couple were to have a large family, list all the possible types to which the children could belong.

6.5. Three alleles at a certain locus in mice control the intensity of color such that *D* is dominant and produces full color, *d* is recessive to *D* and results in partial dilution of color, and a third allele, d^1, is recessive to both *D* and *d* and results in embryonic death when homozygous. **(a)** Assume that a mouse of the genotype *Dd* is mated to one of genotype dd^1; list the possible genotypes and phenotypes that could be produced. **(b)** Suppose that all possible crosses (four) were made between full-color mice and diluted-color mice of the types produced in part (a); what phenotype ratio would be expected among the progeny?

REFERENCES

BELYAEU, D. K., L. N. TRUT, and A. O. RUVINSKEY. 1975. Genetics of the W-locus in foxes and expression of its lethal effects. *Journal of Heredity* 66 : 331.

EVANS, J. W., A. BORTON, H. F. HINTZ, and L. D. VAN VLECK. 1977. *The Horse.* W. H. Freeman and Company, Publishers, San Francisco. Chapter 15.

GREEN, E. L., ed. 1968. *Biology of the Laboratory Mouse.* Dover Publications, Inc., New York. Chapter 21.

OLSON, T. A. 1981. The genetic basis for piebald patterns in cattle. *Journal of Heredity* 72 : 113.

SRB, A. M., R. D. OWEN, and R. S. EDGAR. 1965. *General Genetics*, 2nd ed. W.H. Freeman and Company, Publishers, San Francisco. (Plumage color in pigeons, p. 263)

STERN, C. 1973. *Principles of Human Genetics*, 3rd ed. W.H. Freeman and Company, Publishers, San Francisco. (Human blood groups)

WARWICK, B. L., and P. B. DUNKLE. 1939. Inheritance of horns in sheep. *Journal of Heredity* 30 : 325.

SEVEN

Gene Frequencies

The concept of gene frequencies refers to the relative abundance of a gene in a population compared to the abundance of the alleles of that gene. For example, if there are two alleles affecting a certain trait in a large group of animals, the percentage of one allele in the total population could vary from zero to 100 percent. The frequency of the second allele would be 100 minus the frequency of the first allele. The relative frequencies of the two alleles in percentage would have to total 100 percent. Frequencies of alleles are quite often expressed in decimal fractions ranging from zero to 1.0. The sum of frequencies of all alleles in a system in this case must equal 1.0.

In this chapter we deal with the importance and applications of gene frequencies, the relationship between gene frequencies and genotypic frequencies, how to calculate gene frequencies, and some factors that influence changes in gene frequencies. The study of gene frequencies is actually an extension of the study of probabilities that began in Chapter Four. Many of the applications of gene frequency calculations are related to the concepts of probabilities. Knowing the frequency of a detrimental recessive gene in a population can help in determining the likelihood that the gene will occur again, and can provide information as to how many normal-appearing individuals might be carrying the recessive gene. This and other applications will be found in this chapter.

7.1 GENETIC EQUILIBRIUM

7.1.1 Random Mating

If the probability that any one male will mate with any female in a population is the same as for all other males, the concept of random mating exists. Random mating is accomplished when the probability of mating any given male to a particular female is directly proportional to the number of females in the population. Random mating can exist for any given trait independently of another. Among a certain population of people, for example, brown-haired men tend to choose blonde-haired women to be their wives. In this instance, selective or assortive mating will exist, not random mating. On the other hand, the same group of men may not have any preference whatsoever for the blood type of their intended mate. With respect to genes for blood type, random mating will have occurred.

Assume that among a very large herd of cattle grazing on open range, one-third are red and two-thirds are black. Further assume that the same ratio of colors exists among both bulls and cows. If random mating is to occur, the probability that a red bull will be mated to a red cow is $\frac{1}{3} \times \frac{1}{3}$, or $\frac{1}{9}$. Table 7.1 lists all the possible combinations of matings and their respective probabilities with random mating.

Carrying the previous example a bit further, suppose that $\frac{1}{2}$ the cows and bulls are horned and $\frac{1}{2}$ are polled. Let us further assume that horned bulls tend to exert some degree of social dominance over polled bulls such that a horned bull is more likely to breed a given cow than is a polled bull. In this instance, there is a degree of assortive or selective mating taking place. Calculations of the probabilities of certain matings could not be made as accurately as in the example with color. Relatively accurate probabilities could be obtained, however, by observing the breeding habits of horned bulls relative to polled ones.

Breeders can interfere with the random mating process by deciding

TABLE 7.1   Probable Crosses within a Random Mating Herd of Cattle Where One-Third Are Red and Two-Thirds Are Black

Color of Bull	Color of Cow	Mating Probability
Red	Red	$\frac{1}{3} \times \frac{1}{3} = \frac{1}{9}$
Red	Black	$\frac{1}{3} \times \frac{2}{3} = \frac{2}{9}$
Black	Red	$\frac{2}{3} \times \frac{1}{3} = \frac{2}{9}$
Black	Black	$\frac{2}{3} \times \frac{2}{3} = \frac{4}{9}$
		$\frac{9}{9} = 1.0$

that certain bulls with regard to color or horned condition, or any other trait, will be mated to certain cows. Some examples of this will be presented later.

7.1.2 The Hardy–Weinberg Rule

This rule, or mathematical concept, provides us with an explanation of how genes are distributed within a population. The concept was described and published independently by the British physician W. Weinberg and the British mathematician G. H. Hardy in 1908. For the rule to be applied with accuracy, there must be no mutations, no migration, and no selection with respect to the particular locus or loci in question. The absence of selection, either by nature or humans, is another way of saying that matings are at random.

A good way to explain the rule is to use an example. Let us suppose that in a large herd of Shorthorn cattle, random mating is being practiced with regard to the red, roan, and white colors. Remember that this trait is affected by two genes with codominant effects. We shall refer to one of the genes as the red gene and use the symbol R, and the other as the white gene using the symbol r.

Without knowing the actual frequencies of the two alleles, a formula can be derived that could be used with any set of frequencies. Let us use the letters p and q to represent the unknown frequencies of the gene for red and the gene for white, respectively. Assuming that fertilization will occur, the probability that a sperm cell carrying a red gene will fertilize an ovum carrying a red gene is calculated as the product of the two separate probabilities. Therefore, the probability that the red genotype (RR) would exist is calculated as the product of the probability of the existence of a red gene in a sperm cell, which is p, and the probability that a red gene exists in an ovum, which is also p. Since, as yet, the value of p is unknown, the probability that the genotype RR exists is simply $p \times p$, or p^2. The probability that any gamete chosen at random would contain the gene r is q; therefore, the probability that two such gametes would unite at fertilization to produce the genotype rr, and thus the probability of the occurrence of white, is $q \times q$, or q^2.

The heterozygous genotype Rr can come about in either of two ways: an ovum carrying the R gene can be fertilized by a sperm cell carrying the r gene, or an r-bearing ovum can be fertilized by an R-bearing sperm cell. In each of these instances, the probability is $p \times q$. The genotypic and phenotypic outcome of either union is identical to the other, so the total probability is $2 \times p \times q$, or $2pq$. Where p and q are expressed as fractions or decimal fractions, the sum of the probabilities for all three possible genotypes must equal 1.0. The resulting formula is the mathematical expression of the Hardy–Weinberg law and takes the form of

$$p^2 + 2pq + q^2 = 1.0$$

Note that this is the binomial expanded to the second power, $(p + q)^2$.

The Hardy–Weinberg law states that after one generation of random mating in a population where the frequencies of two alleles, in the previous example signified as R and r, are p and q, respectively, the genes will segregate such as to produce $p^2 RR$, $2pqRr$, and q^2rr, provided that the effects of migration and mutation are negligible. Now, let us assume that the values of p and q are known to be 0.7 and 0.3, respectively. The percentages of each genotype that would be expected to exist in a large population can be predicted. The results of those calculations are summarized in Table 7.2.

7.2 CALCULATION OF GENE FREQUENCIES

With but a few exceptions, genes are observed in populations in pairs. With respect to any one locus, each individual carries two genes affecting a common trait. The genes at these homologous loci will be either homozygous or heterozygous, and in the population various proportions of each genotype will occur. Calculating gene or allelic frequencies involves counting the individual genes in the case of codominant inheritance, or estimating their number according to the Hardy–Weinberg law in the case of dominant and recessive inheritance.

7.2.1 Genes Expressing Codominance

Traits that are affected by genes expressing codominance between alleles represent a relatively easy example of calculating gene frequencies. It simply amounts to counting the genes at a particular locus and expressing their numbers as a percentage or proportion of the total numbers of genes at that locus in the population or sample. In a particular breed of chickens, black, blue, and white feather colors represent such an example. The black

TABLE 7.2   Expected Proportions of Red, Roan, and White Cattle in a Random Breeding Herd Where the Frequency of the Red Gene Is 0.7

Genotypes	Phenotypes	Proportions
RR	Red	$p^2 = 0.7^2 = 0.49 =$ 49%
Rr	Roan	$2pq = 2(0.7)(0.3) = 0.42 =$ 42%
rr	White	$q^2 = 0.3^2 = 0.09 =$ 9%
		1.0 100%

and white phenotypes are produced by homozygous genotypes, while blue feather color is the result of the heterozygous state of two codominant alleles.

Suppose that within a flock of 150 chickens, 95 are black, 50 are blue, and the remaining 5 are white. Using the symbols B and b for black and white, respectively, the genotypes for the three colors would be BB (black), Bb (blue), and bb (white). Since each color can be identified as to how many of each of the two genes are present, calculating their relative frequencies becomes a matter of simple counting the genes. Each black chicken carries two genes for black ($2 \times 95 = 190$), and each blue chicken carries one gene for black (50). The total number of black genes in the flock would be $190 + 50 = 240$. The total of all genes affecting these colors would be twice the number of birds in the flock ($2 \times 150 = 300$). The frequency of the gene for black is calculated as $240/300 = 0.8$, or 80 percent. The frequency of the gene for white would be calculated as 1 minus the frequency of the gene for black, $1.0 - 0.8 = 0.2$, or 20 percent. This procedure applies regardless of whether or not there has been random mating.

The question may now arise as to whether random mating had been practiced. Given the allelic frequencies that were calculated according to the procedure above, it can be determined using the Hardy–Weinberg formula whether this flock has been breeding at random for this set of traits. Letting $p = 0.8$, the frequency of the gene for black, and $q = 0.2$, the frequency of the gene for white, the numbers of each genotype that should exist with random breeding can be calculated. The number of black chickens expected with random mating is estimated as $150p^2 = 150(0.8)^2 = 150(0.64) = 96$. The number of blue chickens expected with random mating would have been $150 \times 2pq = 150(2)(0.8)(0.2) = 48$. Finally, the number of white chickens expected would have been $150(0.2)^2$, or 6.

In this example, most observers would probably conclude intuitively that the numbers of the three colors observed—95 black, 50 blue, and 5 white—are close enough to the predicted numbers of 96, 48, and 6, respectively, that random mating had indeed been practiced. The deviations between the observed and calculated numbers could be evaluated using the chi-square test to determine if they were small enough to have been caused by chance. If the deviations had been greater than what would normally be expected due to chance, the conclusion would be that random breeding had probably not occurred.

7.2.2 Dominant and Recessive Genes

Frequencies of dominant and recessive alleles cannot be determined by the gene-counting method because of the inability to distinguish between the dominant homozygote and the heterozygote. However, the Hardy–

Weinberg formula can be used to estimate the frequencies if random mating can be assumed, and if migration and mutation effects are negligible.

Let us assume that the frequency of some dominant gene in a population is equal to p, and the frequency of the recessive allele is q. The frequency or proportion of the homozygous recessive genotype aa would be q^2, and that of the genotypes expressing the dominant trait (AA and Aa) would be $p^2 + 2pq$. Counting the number of recessive individuals in the herd or population will not account for all the recessive genes since some of them will be present in the heterozygotes. The number of dominant individuals would not be a reflection of the number of dominant genes since some of them carry recessive genes. The proportion of recessive individuals in the population, expressed as a decimal fraction, represents the probability that two gametes containing the recessive gene were united at fertilization (q^2 = proportion of aa). Estimating the frequency, q, of the recessive gene becomes a purely mathematical matter of taking the square root of the frequency of the recessive genotype q^2. Since the sum of p and q must equal 1, the frequency of the dominant gene is $1 - q$.

The principle of calculating frequencies of dominant and recessive genes can best be explained using an illustration. The presence of the white belt in hogs is under the control of a completely dominant gene; the absence of the belt is a recessive trait. Suppose that in a population of 230 hogs, 147 were belted and the other 83 were not. The proportions of the two traits would be 0.639 belted and 0.361 nonbelted. Two mathematical expressions may be formed: $p^2 + 2pq = 0.639$, and $q^2 = 0.361$. The frequencies of the two alleles would be determined by solving for p and q. The second of the two expressions is the easier to solve since it consists of only one unknown factor. The frequency of the recessive gene is equal to the second root of 0.361, or 0.601. The frequency of the belted, or dominant, gene is $1.000 - 0.601 = 0.399$.

The 147 belted hogs include both homozygotes and heterozygotes, and because of complete dominance, the two cannot be distinguished. However, the approximate numbers of each can be calculated using the Hardy-Weinberg formula. The proportion of homozygotes can be represented by p^2, while $2pq$ would represent the proportion of heterozygotes. The number of homozygous belted pigs (BB) is estimated as $230p^2 = 230(0.399) = 36.6$ or 37. The number of heterozygotes can be estimated by solving the expression $230 \times 2pq = 230(2)(0.399)(0.601) = 110.3$ or 110.

The procedure for calculating the frequencies of dominant and recessive alleles can be summarized in the following steps:

1. Count the number of individuals in the population expressing the recessive trait.

2. Express this number as a decimal fraction of the total population.

3. Obtain the second root of the decimal fraction.
4. This is the estimate of the frequency of the recessive gene assuming that random mating has been practiced.
5. The frequency of the dominant gene is 1 minus the frequency of the recessive gene.

7.2.3 Sex-Influenced Traits

The procedure for calculating the frequencies of alleles affecting sex-influenced traits is similar to that for other dominant and recessive genes except that the proportions of recessive individuals must be separated by sex groups. With the red and mahogany colors in Ayrshire cattle, for example, red is recessive in bulls, and mahogany is recessive in cows. Gene frequencies could be determined by obtaining the square root of the proportion of red cattle in the male population, or obtaining the square root of the proportion of mahogany cattle in the female population.

Assume that in a large random breeding herd of Ayrshire cattle, among 200 calves there were 49 mahogany heifers, 51 red heifers, 91 mahogany bulls, and 9 red bulls. Mahogany in heifers is recessive and represents 0.49 of the female population. The frequency of the gene for mahogany is calculated as the second root of 0.49 or 0.7. The frequency of the gene for red is $1.0 - 0.7 = 0.3$. The numbers of bull calves could have as easily been used to determine these frequencies. Red in bulls is recessive and represents 0.09 of the population. The frequency of the gene for red can be estimated as the square root of 0.09 or 0.3. The frequency of the gene for mahogany would therefore be $1.0 - 0.3 = 0.7$. The number of heterozygotes would be expected to be the same for both sex groups, $2pq(100) = 2(0.3)(0.7)(100) = 42$. In the heifers, these 42 calves would be included among the 51 red ones. In the bulls, they would be included among the 91 mahogany ones.

7.2.4 Sex-Linked Genes

The nature of inheritance of genes carried on the X chromosomes in mammals (or the Z chromosomes in birds) presents another variation in the procedure for determining gene frequencies. Since male mammals carry only one X chromosome in each of their cells, they will carry only one gene affecting an X-linked trait. The number of males expressing a given trait expressed as the decimal portion of the total number of all males is equal to the frequency of the gene controlling that trait. Do not obtain the square root of this number. It makes no difference whether the alleles involved are related by dominance and recessiveness or codominance.

Frequencies of genes may be calculated in females using procedures identical to those which have previously been described. The assumption is made that for most genes, the frequencies will be the same in both sexes.

The example of red-green colorblindness in humans will be used to illustrate how to calculate frequencies of X-linked genes. Assume that 1000 men and 1000 women were examined to determine if they were colorblind. Suppose that 61 of the men were found to be colorblind, while only four women were found. The frequency of colorblindness in men is 61/1000 or 0.061. Since each man carries only one gene for the trait, this is also the frequency of the gene. The frequency of the normal gene is 1.000 − 0.061 = 0.939.

If the frequencies of the two alleles are assumed to be the same in men and women, those values can be substituted for p and q in the Hardy–Weinberg formula to estimate the number of colorblind women expected. letting $q = 0.061$, then $q^2 = 0.0037$, the proportion of the female population expected to be colorblind. Among 1000 women this would be 3.7 or approximately 4, the number counted in this illustration. Notice that in order for women to be colorblind, they must carry two recessive genes. As illustrated, the probability that two recessive genes will exist together is considerably less than the probability that one would exist alone.

Among the 996 women with normal vision in the example above, it might be interesting to know how many would be expected to carry a recessive gene. The portion of the Hardy–Weinberg formula representing the incidence of heterozygotes is $2pq$. Substituting the values of p and q, we find that 1000(2)(0.939)(0.061), or 114 of the 1000 women should be heterozygous. Note that the heterozygotes are not identified.

7.2.5 Multiple Alleles

As described in Chapter Six, the concept of multiple alleles refers to a situation where a trait is controlled by three or more alleles. The Hardy–Weinberg formula cannot be used in the form in which it was presented previously; however, the principle involved is essentially the same. To calculate the relative frequencies of three alleles, the trinomial ($p + q + r = 1.0$) must be expanded to the second power. The expansion would produce the expression

$$p^2 + 2pq + 2pr + q^2 + 2qr + r^2 = 1.0$$

Each term in the expansion would represent the proportion of one of the six possible genotypes that could exist in a random-breeding population involving a trait controlled by three alleles. The frequencies of the three alleles would be represented by p, q, and r, respectively.

This concept can be illustrated using three alleles controlling the Hereford spotting pattern, solid color, and the irregular spotting pattern as seen in Holstein cattle. The Hereford (white face) pattern is controlled by a completely dominant gene S^h. The irregular white-spotted pattern is controlled by a completely recessive gene s. The solid pattern as seen in Angus cattle is controlled by an allele S which is completely recessive to S^h, but completely dominant to s. The frequencies of the three alleles will be represented by the letters p, q, and r, respectively. Assume that the values for purposes of this illustration are known to be 0.2, 0.5, and 0.3, respectively. The results of substituting these values into the expanded trinomial are summarized in Table 7.3.

The frequency or probability of the white-face pattern (S^h) is represented by the terms $p^2 + 2pq + 2pr$, the sum of which is 0.36 (0.04 + 0.20 + 0.12 = 0.36). The proportion of solid-colored cattle (SS and Ss) is represented by the sum of the terms $q^2 + 2qr$, which is 0.55. Finally, the homozygous recessive spotted cattle (ss) are represented by the single term r^2, which is 0.09 of the population.

In the previous example, the frequencies of three alleles were known and the frequencies of genotypes and phenotypes were calculated using the expanded trinomial. If the allelic frequencies were not known, they could be calculated if the phenotypic frequencies were known. The process is very similar to using the Hardy–Weinberg formula to help calculate the frequencies of two dominant and recessive alleles in an earlier problem. The first step in the procedure is the same in both cases: determine the number of homozygous recessive individuals in the population, express it as a decimal fraction, and extract the second root; the result is the frequency of the recessive allele.

The ABO blood group system in humans is controlled by three alleles and will serve to illustrate how the frequencies of multiple alleles can be calculated. Three alleles, A, B, and a, control four phenotypes, blood groups A, B, AB, and O. The alleles A and B are completely dominant to

TABLE 7.3   Expected Numbers of Genotypes and Color Patterns
in Cattle When Frequencies of the S^h, S, and s Alleles
Are 0.2, 0.5, and 0.3, Respectively

Phenotypes	Genotypes	Genotypic Frequencies	Genotypic Probabilities	Phenotypic Probabilities	Numbers per 100
White face	S^hS^h	p^2	0.04		
	S^hS	$2pq$	0.20		
	S^hs	$2pr$	0.12	0.36	36
Solid	SS	q^2	0.25		
	Ss	$2qr$	0.30	0.55	55
Spotted	ss	r^2	0.09	0.09	9
			1.00	1.00	100

a, but express codominance between themselves. The genotypes *AA* and *Aa* produce blood group A; blood group B is the product of either genotype *BB* or *Ba*; the genotype *AB* produces the blood group designated as AB; and finally, the homozygous recessive *aa* produces blood group O. Let the frequencies of the alleles *A*, *B*, and *a* be represented by the letters *p*, *q*, and *r*, respectively. Assume that in a randomly selected sample of 100 students, there were 49 with group O blood, 32 with group A, 15 with group B, and only 4 with group AB. The information is now available to calculate, or estimate, the frequencies of the three alleles.

The frequency or proportion of the homozygous genotype *aa* (group O individuals) is r^2, or 0.49. The frequency of the recessive gene *a* is estimated as the square root of 0.49, or 0.7. The proportion of the population represented by *a* and just one of the other alleles can now be used to estimate the frequency of the second allele. The equation $p^2 + 2pr + r^2$ represents that part of the sample population which are groups A and O, which is $0.32 + 0.49 = 0.81$. Since $p^2 + 2pr + r^2 = (p + r)^2$, then $(p + r)^2 = 0.81$, and $p + r = 0.9$, the second root of 0.81. The value of *r* has been calculated as 0.7; therefore $p = 0.9 - 0.7 = 0.2$. The sum of $p + q + r$ must equal 1.0, and *p* and *r* are now known, so $q = 1.0 - (0.7 + 0.2) = 0.1$.

7.3 FACTORS AFFECTING CHANGES IN GENE FREQUENCIES

The relative frequencies that exist between and among alleles in a population are not fixed but are subject to changes during each generation. Several factors can cause such change. These include the effects of natural and artificial selection, mutation, chance, migration and isolation, and mixture of populations.

7.3.1 Selection

Selection is a process that involves forces or factors that determine which animals in a given generation will breed and produce the next generation. These forces may be of natural origin, in which case it is called *natural selection*. If the forces involved in the selection process are under human control, the process is called *artificial selection*. In either case, the genetic effect is the same; the frequency of one gene is changed relative to its allele or alleles. The process of change in gene frequencies normally moves at a faster rate with artificial selection than with natural selection.

To illustrate how selection can change gene frequencies, let us look at the horned-polled trait in a population of several thousand cattle. Assume that this is a domestic population where the frequency of the horned and polled alleles are 0.5 each, and where there has been a total disregard for whether cattle were horned or polled. In other words, there has been random mating for this trait. Recall that the gene for the polled condition is completely dominant to the gene for horns. With the gene frequencies given, approximately three-fourths of the population should be polled while one-fourth should be horned. Among the 75 percent of the cattle that are horned, one-third of them should be homozygous and two-thirds should be heterozygous.

If a program is begun where only polled cattle will be used for breeding purposes, the frequency of the gene for polled will increase, while the frequency of the gene for horns will decrease. Since all horned cattle will be culled from the breeding herd, the only recessive genes remaining are those that exist among the heterozygotes. The frequency of the recessive gene in the new herd can be estimated by the "gene-counting" method used for codominant alleles. Among the 75 percent of the original herd that was polled, 50 were estimated to be heterozygous, so among 75 polled cattle there should be 50 recessive genes. The total number of genes at this locus in 75 cattle is 150; therefore, the frequency of the recessive gene is now $\frac{1}{3}$ or 0.333, while the frequency of the dominant gene is $\frac{2}{3}$.

Among the calves produced from this group of polled cattle, about $\frac{1}{9}$ of them $(\frac{1}{3})^2$ should develop horns. Using a similar procedure of culling only the horned cattle in each generation, the new frequency of the horned gene would be 0.25, and the percentage of calves expected to develop horns in the next generation should be 6.25 ($0.25^2 \times 100$). The A-curves in Figure 7.1 illustrate the changes in frequencies of dominant and recessive alleles over several generations when selection is against the recessive homozygote.

The rate of change in gene frequencies can be increased if more genes can be culled during any one generation. For example, in the illustration above, if one heterozygous parent were culled for every recessive homozygote culled, the rate of change would be as shown in Figure 7.1 by the B-curves. By this principle, it can be concluded that if both heterozygous parents could be culled along with each recessive homozygote, the changes in gene frequencies would occur at an even greater rate.

7.3.2 Mutation

A *mutation* represents a chemical change in a gene such that it functions differently than before the change. It controls the synthesis of a different gene product or enzyme. The mutated gene continues to affect the same trait and resides at the same locus; therefore, it is an allele of the original gene. If a gene mutates at a constant rate, the frequency of this gene will decrease slightly while the frequency of the mutant allele will increase slightly. The rate at which genes mutate is quite variable, but is usually

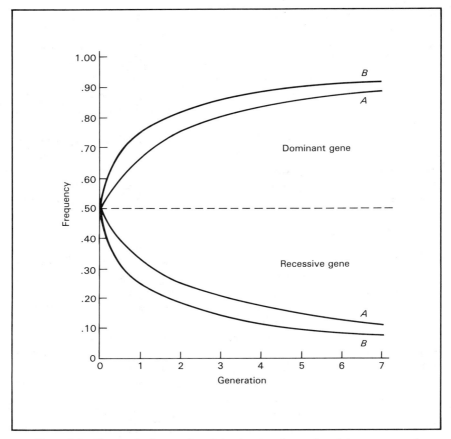

Figure 7.1 Changes in frequencies of dominant and recessive alleles over several generations when selection is against the recessive allele. Curve A: elimination of all recessive homozygotes each generation; curve B: elimination of all recessive homozygotes and one heterozygous parent each generation.

relatively low, possibly in the range of one mutation for each million or more replications of a gene.

To illustrate the relationship between selection and mutation and their effect on gene frequencies, let us look at the frequency of the gene for red color in the Black Angus breed of cattle. The frequency of the gene for red has been estimated to be between 0.05 and 0.08. With one generation of random mating for color, this would result in the birth of between 25 and 64 red calves per 10,000 cattle. Since red calves and some of their heterozygous parents would very likely be culled, it would seem that the frequency of the red gene should continue to decline. One explanation for why it is apparently not decreasing is that the mutation rate from black to red genes may be high enough to produce an equilibrium between selection for the red gene and the mutation rate for red. Another probable explanation is that the selection process among Angus cattle slightly favors the heterozygote for the black and red alleles over homozygous dominant individuals.

7.3.3 Genetic Drift, Migration, and Isolation

These three factors are discussed together, due to some similarities and interrelationships among them. In a particular population the frequencies of alleles will vary from generation to generation due to chance and the sampling nature of inheritance. This random shift or fluctuation in frequencies is referred to by some geneticists as *genetic drift* or *random genetic drift*. Greater shifts are likely to be observed when small groups or samples of a population are removed for breeding purposes. If these samples are then isolated from the larger population, frequencies may become established in the new population that are quite different from those of the parent population.

When a small group or sample is removed from a larger population, this can be thought of as migration. In human populations, a small group may leave their native home or country and move to a new country or location. This is one of the usual examples of migration. Many examples of migration can be cited among wild animals, too. Due to chance, the frequencies of various genes in the small migrating population may differ from those of the native or parent population. The new population may be geographically isolated, or they may choose to be genetically isolated by refusing to mate or intermarry with members of the local population. Inbreeding within a small population has an effect similar to isolation.

It should be noted that with respect to genetic drift, migration, and isolation, any changes in gene frequencies that occur are due to chance and not specifically to migration and isolation. The effect of migration and isolation is to allow genotypes to become established according to the new gene frequencies that were caused by the sampling nature of inheritance.

7.3.4 Mixture of Populations

Change in frequencies of certain genes may be brought about by mixing two populations with different gene frequencies. The frequency of any gene in the new population is essentially the mean of the frequencies in the two beginning populations. Suppose that a farmer has 150 cattle, where the frequency of the gene for horns is 0.95. This means that if random mating has been practiced, about 90 percent of the cattle are horned. Assume that six new bulls are purchased, one of which is horned, two are heterozygous, and three are homozygous dominant. The frequency of the horned gene among these bulls is 0.333. Assuming that these are the only bulls in the herd, after one generation of random mating, the new frequency would be about 0.64.

7.3.5 Inbreeding and Outbreeding

Inbreeding and outbreeding are mating systems the definitions of which depend on the genetic relationship of the animals being mated. Genetic relationship between two animals refers to the percentage of genes that they have in common. *Inbreeding* is the mating of animals more closely related than the average of the population; *outbreeding* is the mating of animals less closely related than the average of the population.

Inbreeding can be thought of as a form of genetic isolation. If a population is truly isolated, inbreeding will tend to occur due to the limited options to mate with unrelated animals. If inbreeding occurs among a group of animals not geographically isolated, the effect is the same; therefore, inbreeding is a kind of artificial isolation. The inbreeding itself will not change the frequencies of genes, even as isolation does not. The initial frequencies of alleles will have been determined when the population began inbreeding. Any subsequent changes in frequencies of genes will occur due to selection, mutation, or random sampling effects.

If outbreeding is performed in a population with approximately equal sex ratio and the gene frequencies at a given locus are the same in both sexes, the frequencies of the alleles should not change, due directly to outbreeding.

STUDY QUESTIONS AND EXERCISES

7.1. Define each term.

assortive mating	inbreeding	outbreeding
binomial	isolation	random mating
genetic drift	migration	selection
genetic relationship	mutation	trinomial

7.2. Explain the purpose and function of the Hardy–Weinberg rule.

7.3. Why is it not necessary to use the Hardy–Weinberg formula to calculate the frequencies of codominant alleles?

7.4. By way of discussion rather than calculation, explain how to determine the frequency of a recessive gene in a population.

7.5. Why is it not necessary to use the Hardy–Weinberg formula to calculate the frequencies of sex-linked genes?

7.6. Explain how selection can change gene frequencies.

7.7. How do mutations change gene frequencies?

7.8. Explain how changes in gene frequencies due to selection can be offset by changes due to mutations.

7.9. How can changes in gene frequencies due to selection and mutations work together to increase the frequency of a given gene?

7.10. What is meant by the sampling nature of inheritance, and how can it result in changes in gene frequencies?

7.11. The frequency of the recessive gene for red color in Angus cattle is about 0.065, while the frequency of the recessive gene for horns is essentially zero. Intensive selection has been practiced against both genes for many years. Provide some plausible reasons why the frequency of the red gene is not lower.

7.12. Explain why migration and isolation in themselves do not change the frequencies of genes.

7.13. How are the effects of isolation and inbreeding similar?

7.14. How may the effects of mixing population and outbreeding be similar?

7.15. Explain why the mixing of populations is not necessarily the same as outbreeding.

7.16. What effect do inbreeding and outbreeding have on gene frequencies?

PROBLEMS

7.1. Black color in Berkshire hogs is controlled by a gene that is codominant with its allele for red color. The heterozygote is red with varying amount of black spotting. **(a)** Assume that in a herd of 100 hogs, 35 are black, 21 are red, and the remaining 44 are spotted. What are the frequencies for the genes for red and black? **(b)** Suppose that a small group of hogs were separated from the larger herd and allowed to breed among themselves at random. What proportions of the three colors would be expected in the litters if the new frequency of the gene for black were 0.3? **(c)** If a random sample of sows from the herd in part (a) were bred to black boars, what proportions of the three colors would be expected in the litters of pigs?

7.2. The palomino, chestnut, and white colors in horses are controlled by a pair of codominant alleles such that the genotypes for white and chestnut are homozygous, and palomino is heterozygous. **(a)** Suppose that in a large herd of horses, 47 percent are chestnut, 51 percent are palomino, and 2 percent are white; what is the frequency of the chestnut gene? **(b)** Based on the frequencies calculated in part (a), determine the percentages of the three colors that should have been expected if breeding had been completely random. **(c)** What conclusion can be made as to whether there has been random mating?

7.3. Platinum color in foxes is produced by a heterozygous genotype involving interactions between a gene for silver and a codominant lethal gene. The silver gene in the homozygous state produces silver color; the lethal homozygote usually dies during early embryonic development. **(a)** Calculate the frequency of the silver gene among a group of foxes where 48 percent are silver and 52 percent are platinum. **(b)** Using the frequencies calculated in part (a), estimate the percentage of the two colors that would be expected in the litters with random mating. **(c)** Calculate the new frequency for the gene for silver. **(d)** Provide an explanation for the change in frequencies.

7.4. The rose comb trait in chickens is dominant to the usual single comb. Suppose that in a flock where random mating has been practiced with respect to comb type, 5 percent of the birds have rose combs and 95 percent have single combs. **(a)** What is the frequency of the rose-comb gene in this flock? **(b)** If there were 100 chickens in the flock, how many of the five rose-combed birds would be expected to be heterozygous? **(c)** About how large would the flock have to be to expect at least one homozygous dominant rose-combed bird?

7.5. A condition in cattle known as "snorter" dwarfism is caused by a recessive gene when in the homozygous state. Normal cattle with regard to size possess at least one dominant gene. **(a)** In a large herd that produces about 650 calves each year, over a three-year period dwarf calves were produced at the rate of 23 calves per year. Calculate the frequency of the dwarf gene. **(b)** Among 800 cows in the herd, how many would be expected to be "carriers"?

7.6. In some polled cattle a shapeless, horny growth develops where the horns would normally be. This condition is called scurs. It is apparently controlled by a pair of sex-influenced alleles such that the gene for scurs is dominant in bulls and recessive in cows. The inheritance is much the same as for the red and mahogany colors in Ayrshire cattle described in Chapter Five. Assume that in a herd of 236 polled cows, only four were scurred. **(a)** What is the frequency of the scur gene among the cows? **(b)** How many of the 236 cows would be expected to be heterozygous? **(c)** If the next crop of calves included 100 bull calves, how many would be expected eventually to develop scurs? **(d)** If the 236 cows were bred to nonscurred bulls, is it possible for calves to be produced that would develop scurs? Explain.

7.7. A certain kind of colorblindness in humans is controlled by a sex-linked recessive gene. A survey of 1000 men showed that 32 of them were colorblind. **(a)** What is the frequency of the colorblind gene? **(b)** If 1000 women were tested for colorblindness, about how many would be expected to be colorblind? **(c)** How many of 1000 women would be expected to be heterozygous carriers of the recessive gene?

7.8. In female cats, two codominant alleles produce the colors yellow, black (non-yellow), and tortoiseshell. Yellow and black are homozygous, while tortoiseshell is heterozygous. Among 338 female cats in a study of cat colors in London, there were 7 yellow ones, 54 tortoiseshell, and 277 that were neither yellow nor tortoiseshell. Among 353 male cats, 42 were yellow and 311 were not yellow or tortoiseshell. **(a)** Calculate the frequency of the gene for yellow color in this population. **(b)** Using the frequencies calculated in part (a), determine the theoretical numbers of each color that would have been expected among female cats with random mating. **(c)** Calculate the numbers of male cats that would have been expected of each color. **(d)** How do the theoretical numbers compare to the observed numbers?

7.9. The ABO blood group system in humans is controlled by three alleles, A, B, and a. The alleles A and B express codominance between them and are completely dominant to allele a. A detailed description of this trait is found in Chapter Six. Six genotypes are possible, producing four phenotypes: AA and Aa produce blood group A, BB and Ba produce blood group B, AB produces group AB, and aa produces group O. **(a)** In a study of approximately 50,000 people, it was found that 42.3 percent were in group A, 9.4 percent were in group B, 3.5 percent were group AB, and 44.8 percent were in group O. Calculate the frequencies of the three alleles. **(b)** In a sample of people in Moscow, USSR, frequencies of the A, B, and a alleles were known to be 0.261, 0.154, and 0.585, respectively. Calculate the percentage of the population expected for each of the four blood groups.

REFERENCES

BURNS, G. W. 1980. *The Science of Genetics*, Chapter 14, *Population Genetics*. Macmillan Publishing Co., Inc., New York.

SEARLE, A. G. 1949. Gene frequencies in London cats. *Journal of Genetics* 49 : 214.

SEARLE, A. G. 1959. Gene frequencies in Singapore cats. *Journal of Genetics* 56 : 111.

STERN, C. 1943. The Hardy–Weinberg Law. *Science* 97 : 137.

Epistasis and Modifying Genes

Epistasis involves the interaction between genes that are not alleles. To illustrate epistasis, at least two pairs of genes must be used. Genes at one locus interact with genes at another locus to produce phenotypes that cannot be produced by either locus acting independently. Epistatic interactions may also be defined as a situation where genes at one locus modify the expression of genes at a second locus. The definition of epistasis is somewhat analogous to that of dominance, except that dominance involves the interaction between alleles.

8.1 PRINCIPLES OF EPISTASIS

8.1.1 Mechanism of Action of Epistatic Genes

Most phenotypes are the results of a series of chemical reactions controlled by enzymes. The production of these enzymes is controlled by genes. A modification of the melanin reaction described in Chapter One can be used to provide one simple illustration of how epistasis works:

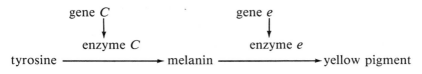

The basic pigment melanin is produced in part due to the enzymatic actions of gene *C*. The melanin is further modified by the enzyme whose synthesis is controlled by gene *e* to produce yellow pigment. If gene *C* is not present, as in the genotype *cc*, the enzyme that catalyzes the production of melanin will be absent; therefore, melanin will not be synthesized. So, even if gene *e* is present, in the absence of melanin, the yellow pigment cannot be produced. Since it is the recessive gene at the *C*-locus that does not allow melanin to be produced, it is the epistatic gene. Literally, the word *epistasis* means to "stand upon." In the genetic context, one gene suppresses the expression of another. The phenotype being suppressed or modified is referred to as the *hypostatic trait*, a term that literally means to "stand below."

8.1.2 Classical Dihybrid Ratios

A knowledge of *dihybrid ratios* is important to the study of epistasis. It helps in deciding whether or not epistasis is present, and if it is, which kind of epistasis is involved. *Classical ratios* are those we have come to expect from monohybrid and dihybrid crosses. Since epistasis must involve at least two pairs of genes, we shall deal only with the classical dihybrid ratios. When the homozygotes *AABB* and *aabb* are crossed (P_1 cross) to produce the dihybrid F_1 (*AaBb*), the mating of these dihybrids among themselves produces the following classical F_2 genotypic ratio:

$$
\begin{array}{ll}
1 & AABB \\
2 & AABb \\
1 & AAbb \\
2 & AaBB \\
4 & AaBb \\
2 & Aabb \\
1 & aaBB \\
2 & aaBb \\
1 & aabb \\
\end{array}
$$

Depending on the type of interaction between the alleles at each locus, there could be any of several possible phenotypic ratios among these F_2 individuals. If both pairs of alleles expressed codominance, we would expect to see nine different F_2 phenotypes in the ratio $1 : 2 : 1 : 2 : 4 : 2 : 1 : 2 : 1$, the same as the genotypic ratio. If the alleles at one locus expressed codominance and those at the other locus expressed complete dominance and recessiveness, the expected phenotypic ratio would be $3 : 6 : 3 : 1 : 2 : 1$. Further, if the alleles at both loci were controlled by complete dominance and recessiveness, the expected phenotypic ratio would be the now familiar 9 :

3 : 3 : 1. These classical ratios are illustrated in Table 8.1 on lines 1, 2, and 4, respectively.

When dihybrid ratios are produced such that two or more phenotypic classes of the classic ratios are combined, this is evidence of epistasis. The classic ratios are determined by how the alleles at each locus interact. If the interaction of the alleles at each of two loci is one of complete dominance, a dihybrid ratio of 9 : 3 : 3 : 1 is expected. However, if the actual ratio were observed to be 12 : 3 : 1, for example, epistasis is suggested because at least two of the components of the classical ratio have been combined. The specific kind of epistatic interaction is indicated by the ratio itself.

8.2 SIMPLE EPISTASIS

The simplest examples of epistasis are those where the genes at one locus modify the expression of genes at a second locus, but the genes at the second locus do not modify those at the first. Three examples follow.

8.2.1 Dominant Epistasis

With this kind of epistasis, a dominant gene at one locus suppresses or modifies the expression of genes at a second locus. This example is illustrated on line 8 of Table 8.1. In cats, at least two kinds of white can exist. One kind is the true albino white with pink eyes, which is quite rare. Another kind of white exists that is controlled by a dominant epistatic color inhibiting gene W. However, pigment is inhibited only in the hair and not in the eyes. Black in cats is controlled in part by a dominant gene B, while the chocolate brown results from the recessive allele b. If we assume that a homozygous black male ($BBww$) is bred to a homozygous white cat known to be carrying the homozygous recessive genes for brown ($bbWW$), the F_1 litter should all be white and heterozygous ($BbWw$). If several matings are made among the cats from the F_1 litter (dihybrid crosses), the four classical genotypic groups would be produced, but only three phenotypes would exist:

$$\left. \begin{array}{l} 9\ B\_W\_ \\ 3\ bbW\_ \end{array} \right\} \text{12 white cats}$$

$$\begin{array}{l} 3\ B\_ww \\ 1\ bbww \end{array} \quad \begin{array}{l} \text{3 black cats} \\ \text{1 brown cat} \end{array}$$

The dominant inhibitor gene is epistatic to both genes at the B locus. Because both black and brown are suppressed, they are the hypostatic traits.

TABLE 8.1 Dihybrid Ratios Illustrating Several Kinds of Gene Interaction

Phenotypic Category	Line	Type of Gene Action	1	2	3	4	5	6	7	8	9
			AaBb	AaBB	AABb	AABB	AAbb	Aabb	aaBB	aaBb	aabb
More than four phenotypic classes	1	Codominance for both A and B (no epistasis)	4	2	2	1	1	2	1	2	1
	2	A codominant (no epistasis), complete dominance for B	6			3	1	2	3		1
	3	A codominant and epistatic to B and b	6			4		2	3		1
Four phenotypic classes	4	Complete dominance for A and B (no epistasis), two traits per genotype		9			3			3	1
	5	Complete dominance for A and B (epistasis), one trait per genotype		9			3			3	1

	No.	Type of gene interaction				
Fewer than four phenotypic classes	6	*A* codominant and lethal, *B* completely dominant, *bb* epistatic to *A* and *a*	6	(4) This class missing	3 Incl. *aabb*	3
	7	Recessive epistasis, *aa* epistatic to *B* and *b*	9		3	4
	8	Dominant epistasis, *A* epistatic to *B* and *b*	12		3	1
	9	Dominant and recessive epistasis, *A* epistatic to *B* and *b*, *bb* epistatic to *A* and *a*	13	Includes *aabb* from last column	3	
	10	Duplicate dominant epistasis, *A* epistatic to *B* and *b*, *B* epistatic to *A* and *a*	15			1
	11	Duplicate recessive epistasis, *aa* epistatic to *B* and *b*, *bb* epistatic to *A* and *a*	9			7
	12	Duplicate interaction	9		6	1

Source: Adapted from Snyder and David (1957).

8.2.2 Recessive Epistasis

Line 7 of Table 8.1 lists the theoretical dihybrid ratio associated with this kind of epistasis. The albino gene exists in many species of mammals and represents one of the best examples of a trait controlled by a recessive epistatic gene. A dominant gene *C* controls the production of the pigment melanin. The recessive gene in the homozygous state (*cc*) results in failure of the enzymatic system that produces melanin, thus causing the condition called albinism. In rabbits, as is the situation in many other mammalian species, black hair color is dominant to brown. The dominant and recessive genes can be represented by upper- and lowercase letters *B* and *b*. A dihybrid cross (*BbCc* × *BbCc*) would produce the following four genotypic groups, but only three colors:

$$9\ B\_C\_ \qquad \text{9 black rabbits}$$
$$3\ bbC\_ \qquad \text{3 brown rabbits}$$
$$\left.\begin{array}{l} 3\ B\_cc \\ 1\ bbcc \end{array}\right\} \quad \text{4 albino rabbits}$$

The expression of epistasis is not limited to dihybrid crosses. For example, let us suppose that a dihybrid black rabbit is mated to an albino carrying two brown genes (*bbcc*); this would constitute a dihybrid testcross. The classic dihybrid testcross would be expected to produce the following four genotypes:

$$1\ BbCc \qquad \text{double dominant group}$$
$$1\ bbCc \qquad \text{one dominant, one recessive}$$
$$1\ Bbcc \qquad \text{one recessive, one dominant}$$
$$1\ bbcc \qquad \text{double recessive group}$$

These are the same four groups or classes produced by the dihybrid cross above, but with a different ratio. The actual phenotypic ratio would be determined by noting which two of the groups would be combined due to epistatic effects. In this case, the last two groups, those with the *cc* genotypes, would both express the albino trait. The actual phenotypic ratio would be 1 black : 1 brown : 2 albino.

Referring to the previous example of color inhibition in cats, if a testcross had been performed on dihybrids, the results would have been 2 white : 1 black : 1 brown. The point is that the same two groups that were combined to change the classic 9 : 3 : 3 : 1 ratio to 12 : 3 : 1 would be combined to change the 1 : 1 : 1 : 1 classic testcross ratio to a 2 : 1 : 1.

8.2.3 Codominant Epistasis

In this example, listed in line 3 of Table 8.1, a codominant gene at one locus acts to suppress the expression of genes at a second locus. In the presence of certain other modifying genes in horses, a dominant gene E results in the familiar bay color and pattern. Bay horses have a reddish or brownish-red body with black on the legs, mane, and tail. The recessive gene e in the homozygous state produces a color called chestnut. A more detailed description of the inheritance of these colors will be presented later. A dilution gene c, which is codominant with its allele C, acts in such a way that when heterozygous, bay is diluted to buckskin, and chestnut to palomino. When the dilution genes are present in the homozygous state (cc), the color is almost completely diluted to white, off-white, or a very light cream color called cremello.

Since one set of alleles is controlled by complete dominance and recessiveness, and the other set of alleles by codominance, the classic dihybrid ratio is not $9:3:3:1$, but $3:6:3:1:2:1$. If we were able to observe the progeny from a very large number of dihybrid crosses, the theoretical expected results would be as follows:

3 CCE_-	3 bay
6 CcE_-	6 buckskin
3 ccE_- }	
1 $ccee$ }	4 white or cream (cremello)
2 $Ccee$	2 palomino
1 $CCee$	1 chestnut

8.3 DUPLICATE EPISTASIS

In each of the three examples in Section 8.2, epistasis was acting in only one direction. Genes at one locus were acting epistatically against genes at a second locus, but the genes at the second locus had no effect on those at the first locus. In the following examples, epistasis will be acting in both directions. Genes at two loci will be interacting simultaneously to modify classic dihybrid ratios. In the examples of simple epistasis, only two genotypic groups or classes of the classic ratios were combined. With duplicate epistasis, in all but one example, three classes will be combined, usually resulting in dihybrid ratios with only two phenotypes.

8.3.1 Duplicate Dominant Epistasis

Dominant epistasis exists when a dominant gene at one locus interacts to modify the expression of genes at another locus. When a dominant gene at the second locus is also epistatic to genes at the first locus, this is duplicate dominant epistasis. The presence or absence of feathers on the shanks of the legs of chickens is apparently controlled by this kind of inheritance.

Two sets of dominant and recessive genes can be represented by the letters
F, *f*, *S*, and *s*. The presence of at least one dominant gene from either of
two loci (*F_ss*, *ffS_*, or *F_S_*) will cause feathers to grow on the legs. The
absence of feathers on the legs is the result of the double-recessive genotype
ffss. Feathered shanks usually appear in the Black Langshan breed of chickens. Most other breeds have featherless, or clean, shanks. If purebred Lang-
shans are bred to purebred Plymouth Rocks, the F_1 progeny will be feather-
legged and dihybrid (*FFSS* × *ffss* → *FfSs*). If these F_1 hybrids are allowed
to mate among themselves, about $\frac{15}{16}$ of the F_2 progeny will develop feathers
on their legs, while only about $\frac{1}{16}$ of them will not. This is similar to the
example in line 10 of Table 8.1.

A testcross of a dihybrid feather-shanked bird would produce
progeny $\frac{3}{4}$ of which would develop feathers on their shanks and $\frac{1}{4}$ would not:

> 1 *FfSs*
> 1 *Ffss* } 3 feathered-legged birds
> 1 *ffSs*
>
> 1 *ffss*   1 with no feathers on the legs

8.3.2 Dominant and Recessive Epistasis

An interesting example of duplicate epistasis exists in chickens, where
a dominant gene at one locus and a recessive gene at another locus are
mutually epistatic. The white seen in the White Leghorn breed is caused by
a dominant inhibitor gene *I* such as the one described in dogs in an earlier
example. The recessive genotype *ii* allows normal colors to be expressed in
the feathers as determined by other genes. The white observed in the White
Plymouth Rock breed is caused by a pair of recessive genes *cc*, while the
dominant gene *C* permits color to be expressed. This recessive white is apparently not the same as the true albino since pigment is present in the eyes.
Pure White Leghorns (*IICC*) mated to pure White Plymouth Rocks (*iicc*)
produce white F_1 birds with the genotype *IiCc*. Allowing these F_1 dihybrids
to mate among themselves would be expected to produce the nine different
genotypes commonly produced from any dihybrid cross. The theoretical
phenotypic ratio usually observed is $\frac{13}{16}$ white and $\frac{3}{16}$ colored. Any bird having
either of the genotypes *I_* or *cc* would be white. These would include *I_C_*,
I_cc, and *iicc*. The only genotypic class that would not be white is the *iiC_*
group. This example is listed in line 9 of Table 8.1.

8.3.3 Duplicate Recessive Epistasis

A third locus controlling white feathers in chickens apparently exists.
A pair of recessive genes (*aa*) produces white feathers as observed in the

White Silkie breed. The dominant gene A allows color to be produced. The action of the genes at the A locus is similar to that of the genes at the C locus described in the previous example. When homozygous White Silkies were crossed with homozygous White Plymouth Rocks, colored F_1 birds were produced. When large numbers of F_2 birds were produced from crosses among the colored F_1 birds, they segregated in a ratio of approximately 9 colored to 7 white. The presence of recessive genes at either the A locus or the C locus produced white:

$$
\begin{array}{ll}
9\ A\_C\_ & 9\ \text{colored} \\
\left.\begin{array}{l} 3\ A\_cc \\ 3\ aaC\_ \\ 1\ aacc \end{array}\right\} & 7\ \text{white}
\end{array}
$$

This type of epistasis is listed in line 11 of Table 8.1.

8.3.4 Duplicate Interaction

The term used to describe this type of epistasis is not as descriptive as some of the previous examples. Dominant genes at two different loci inter- act with recessive genes at opposite loci to produce a common phenotype. This kind of epistasis will be explained using an illustration from hogs. It is referred to in line 12 of Table 8.1. The standard red color in hogs can be expressed only if a dominant gene from each of two different loci (desig- nated the R and S loci) is present along with the regular genes for red. The abbreviated genotype for red would be $R\_S\_$. Double-recessive individ- uals ($rrss$) appear to be nearly white, or a very light yellow. Hogs of geno- types R_ss or $rrS\_$ express a color intermediate to the deep red of purebred Durocs and the near-white of the double recessive, and is referred to as sandy.

Without knowing the physiological effects of the genes involved, it is difficult to define the actual kind of epistasis. It appears to be a case where dominant genes from each of two loci produce red; dominant genes from either, but not both, of two loci produce sandy; and the absence of any dominant genes produces white. The important factor to keep in mind is that according to the definition that has been given, this is an example of epistasis. The classic 9 : 3 : 3 : 1 dihybrid ratio becomes modified to 9 : 6 : 1. It should also serve to illustrate that some examples of epistasis are more complex than others, and that epistasis could well play a more important part in the breeding of domestic animals than may have formerly been thought.

8.3.5 Lethal Genes and Epistasis

This example is presented to illustrate the further complexity of gene interactions. Albinism in mice is inherited in much the same manner as in rabbits and other mammals. Let the recessive albino gene be represented by the letter c, and the allele that allows color to be expressed by an uppercase C. The genotype cc is epistatic to any genes for color. Two of the alleles at the so-called agouti locu (A^y and A) produce yellow and agouti, respectively. Agouti is the common color of most wild rodents and rabbits. Yellow is produced from the codominant interaction of both alleles (A^yA). $A^y A^y$ is lethal and usually results in death of embryos very early in their development. Consequently, when yellow mice are mated together, progeny are usually produced in the ratio of $\frac{1}{3}$ agouti to $\frac{2}{3}$ yellow.

The classic dihybrid ratio of $9:3:3:1$ expected when two pairs of dominant and recessive genes are involved is modified to $3:6:3:1:2:1$ due to the presence of one pair of codominant genes. This ratio is then modified to $3:6:2:1$ after the death of the four individuals containing the A^yA^y genotype. Further modification of the ratio produces a $3:6:3$ because of recessive epistasis involving the albino genes. The genotypes, phenotypes, and ratios associated with the dihybrid cross are listed below:

3 $C\_ AA$	3 agouti
6 $C\_ A^yA$	6 yellow
3 $C\_ A^yA^y$	
1 ccA^yA^y	4 lethal, no phenotype expressed
2 ccA^yA	
1 $ccAA$	3 albino

8.4 INTERACTIONS THAT DO NOT MODIFY RATIOS

Each of the examples of epistasis presented to this point has been associated with a modification of a classical dihybrid ratio or a dihybrid testcross ratio. However, a number of examples of apparent epistatic interactions exist that do not modify these ratios. One example deals with four coat colors in rats. In a particular experiment with laboratory rats, homozygous yellow rats were mated to homozygous black ones, and all F_1 progeny were gray. When the gray rats were mated among themselves, among a large number of F_2 progeny, four colors were produced in the ratio of $\frac{9}{16}$ gray to $\frac{3}{16}$ black to $\frac{3}{16}$ yellow to $\frac{1}{16}$ cream.

The emergence of the $9:3:3:1$ ratio suggests at least three things about the inheritance of these colors. First, the segregation into sixteenths suggests that two pairs of genes are involved; second, the ratio suggests that dominance and recessiveness exists between each of two pairs of alleles; and third, the ratio further suggests that the F_1 gray rats were dihybrids. From principles presented in an earlier chapter, we know that the 9 in the $9:3:3:1$ ratio is associated with genotypes that include at least one dominant gene from each of two pairs, such as A_-B_-). In this example, it means that at least one dominant gene from the A-locus and at least one dominant gene from the B-locus must be present to produce the gray color.

We also know from previously presented information that the phenotypic class represented by the 1 in a $9:3:3:1$ ratio is produced by a homozygous double recessive genotype ($aabb$). This means that cream color in rats is the result of the combined effects of two pairs of recessive genes. The black and yellow colors are both members of phenotypic classes represented by 3's, so these colors are the results of interactions of at least one dominant gene at one locus and a pair of recessive genes at a second locus, of which there are two possible combinations: A_-bb and aaB_-.

In the illustration as first presented, the genotypes of the homozygous yellow and black rats are $AAbb$ and $aaBB$, respectively. The gray F_1 rats would all be dihybrids, $AaBb$. The F_2 progeny would include all the theoretical numbers of genotypes and phenotypes expected from any dihybrid cross with two pairs of dominant and recessive genes: 9 gray rats (4 $AaBb$, 2 $Aabb$, 2 $aaBb$, and 1 $aabb$), 3 black rats (2 $aaBb$ and 1 $aaBB$), 3 yellow rats (2 $Aabb$ and 1 $AAbb$), and 1 cream rat ($aabb$).

This example is clearly an illustration of interaction between nonalleles, because none of the colors can be explained using just one pair of genes. The exact mode of interaction is not clear, but one possible explanation might be presented. Assume that the production of black pigment is under the control of the dominant gene B, while cream color is the result of the homozygous recessive bb. For these two colors to be expressed, recessive genes must be present at the A locus (B_-aa = black, $bbaa$ = cream). The presence of the dominant A gene would modify black to gray (B_-A_-) and cream to yellow (bbA_-). Some of the colors observed in horses, cattle, and dogs are apparently controlled by similar kinds of interactions.

8.5 COAT COLOR IN MAMMALS

The inheritance of color in mammals provides numerous examples to illustrate epistasis and other types of gene action. Selected examples were presented earlier in this chapter and in previous chapters to illustrate mechanisms of specific kinds of gene action. Information on the next few pages summarizes some of the earlier material and also includes some new material. It is presented to provide the reader with a better understanding and appreciation of color inheritance in a few selected species of domestic animals. It should also suggest that coat color inheritance, as well as inheritance of other kinds of traits, is not as straightforward or simple as some earlier examples might have implied.

8.5.1 General Characteristics

The source of all color in animals is melanin. Two basic kinds of melanin exist in most mammals. One form is black melanin (properly called eumelanin); the other is red melanin, or phaeomelanin. The various types and shades of colors observed are produced by modifications of these two pigments. Color in the hair and skin is determined by genes at several loci that affect the synthesis of pigments and associated enzymes, and the distribution and location of pigment granules in the skin cells, hair follicles, and hair shafts.

Considerable amounts of research have been conducted on the inheritance of color in mice, rabbits, dogs, cats, and horses. Information concerning many colors is incomplete, and authorities sometimes are in disagreement as to modes of inheritance. Some interesting similarities in mechanisms of inheritance among various species have been discovered. For this reason, information from one species is sometimes helpful in working out details of inheritance in another.

In the study of coat color inheritance, it is common to begin with the color in wild species. The wild color in many animals is known as agouti, obtained from a wild rodent of that name. This is the familiar color observed in wild rabbits, mice, and rats. The gene for agouti is usually dominant (A_-), while the gene for nonagouti is recessive (aa). Nonagouti is usually black unless modified by other genes. Depending on the species concerned, other alleles may exist in the agouti series.

The major precurser of melanin is the amino acid tyrosine, as explained earlier. One of the enzymes involved in this conversion is tyrosinase; its production is under the control of a gene represented by an uppercase C. The genotype CC controls the production of sufficient enzyme so that enough pigment can be produced for the intense expression of color (full color). A recessive allele at this locus in many mammals results in the complete absence of tyrosinase production and a total deficiency of melanin, the result being the true pink-eyed albino. In some species, other alleles may exist at this locus, often called the albino locus, that act to limit the amounts of tyrosinase production. Two examples of phenotypes resulting from other alleles in the albino series are chinchilla and himalayan in mice and rabbits.

Chinchilla is a pearly gray color, sort of a diluted agouti, taken from the name of an animal of that color. The himalayan pattern is white with colored extremities (see Figure 8.1). More complete lists of alleles located at the albino loci in mice and rabbits were presented in Chapter Six.

Several other loci carry genes which seem to have homologous, but not necessarily identical, effects in different species. Genes at the *B* locus determine whether melanin will be black or brown, or in some cases, depending on other modifiers, red or yellow. Genes at the *E* locus are called the *extension series* of alleles. They are so called because different alleles control or "extend" the relative amounts of eumelanin (black or brown) and phaeomelanin (red or yellow) in the coat. The dominant end of the series is usually all black, while the recessive end is all red or yellow.

A gene at the *D* locus (dilution) may cause clumping of pigment granules, leading to a decrease in light absorption and a dilution of color. This is in contrast to the dilutions that occur because of a reduction in quantity of enzyme and pigment by some genes in the albino series. In some species the dilution gene is recessive; in others it is apparently dominant. A number of genes at several loci control the expression of white from total inhibition of white in the hair and skin to white spotting to interspersion of white hairs among colored hairs (roaning and greying).

8.5.2 Color in Horses

Genes at the *A* locus in horses determine whether black pigment will be fully expressed (as in *aa*), or restricted to the mane, tail, and legs (as in *A_*). The restricted pattern—black mane, tail, and legs with a red body—is called bay (Figure 8.2), and is dominant to black. Some have suggested the existence of a wild-type allele, A^t, dominant to *A* and *a*, to explain the baylike appearance of the wild Przewalski's horse. Whether this allele exists or not is probably not important to the modern horse breeder. It has been

Figure 8.1 The Himalayan pattern in rabbits showing white on the body and black (or other dark color) on the extremities. (Owner: Dian Barker, Babak Kennels, Springfield, MO.)

Figure 8.2   Restricted pattern observed in bays, buckskins, grullos, and some duns. It is characterized by black or brown on the mane, tail, and lower legs, with a lighter color on the head and body.

suggested that seal brown may be controlled by another allele at the A locus, a^t, that is recessive to A and dominant to a. Seal brown appears to be a modification of black with lighter brown areas over the muzzle, over the eyes, inside the legs, and on the flanks. Another source suggests that seal brown may be controlled by a dominant gene at yet another locus.

Genes at the B locus apparently determine whether melanin will be black or brown. A dominant gene B allows for the production of black, except as restricted by $A\_$. The homozygous recessive bb results in the production of chocolate brown. The type of melanin is further affected by genes at the E locus. In most horses, the expression of black or brown requires the presence of the gene E. The homozygous recessive ee limits the expression of black and brown so that phaeomelanin is produced over the whole body, including the mane, tail, and legs. The resulting reddish color, or any of several lighter or darker shades is called chestnut. Depending on the breed, lighter shades are often called sorrel, while the darkest shades are sometimes called liver chestnut.

So far, genotypes for at least five colors in horses have been presented. Black is $aaB\_E\_$; chocolate brown is $aabbE\_$. Bay with black points results from the genotype $A\_B\_E\_$. The genotype $A\_bbE\_$ is probably a bay with brown points. All genotypes carrying ee (A_B_ee, A_bbee, aaB_ee, and $aabbee$) have been grouped together and called chestnut. Although a number of other genes probably control the several shades of chestnut, it seems logical to expect differences in genotypes at the A and B loci to have some effects.

Since black (aa) is recessive to bay (A_-), it would seem that black $\times$ black matings should always produce black foals. In a small percentage of such crosses, bay foals are produced. This has been explained by the existence of a third allele at the E locus. It is dominant to E and e, and has been assigned the symbol E^d. The resulting color is referred to as dominant black or jet black, and is epistatic to A_-. The genotype A_-E^d would show jet black. If it were mated to any black carrying an E allele, a bay foal could result.

The C locus in mice and rabbits is called the albino locus. This reference in horses is probably a misnomer, since the albino gene apparently does not exist, or if it does is lethal. At any rate, the occurrence of pink-eyed albino horses has not been documented. The only known allele of C to exist in horses is the codominant cream dilution gene. Some sources refer to this gene with the symbol c^{cr}, but since it is the only allele of the full-color gene, it is referred to here simply with lowercase c. It is partially epistatic to ee.

Genotypes homozygous for CC will express colors as determined by the genes at the A, B, and E loci as noted above. One dilution gene (Cc) results in partial dilution of phaeomelanin, apparently not affecting eumelanin. The reddish body colors of bays are diluted to buckskin, while the black hair of the mane, tail, and legs is not affected. Chestnuts and sorrels are diluted to palominos. Another characteristic of a palomino is that the mane and tail are always lighter than the rest of the body. Cc apparently has little or no effect on aaB_-E_- (black). The cream dilution gene in the homozygous state results in nearly complete dilution of the reddish colors to white, or very pale yellow. The term *cremello* is sometimes used to describe this color.

Another dilution gene, this one a dominant one, called the dun dilution gene, is partially epistatic to genes at the A, B, and E loci. Black genotypes are diluted to a color known by any of several terms: mouse, steelgray, slate, blue dun, or grullo (pronounced "grew-yo"). Very little, if any, dilution occurs in the mane, tail, and legs, so they are considerably darker than the body. Bay is diluted to dun, which usually has a dark dorsal stripe, sometimes called lineback. The legs, mane, and tail are darker than the body (see Figure 8.2), but often not as dark as in bay and buckskin. Chestnut is diluted to red, orange, apricot, or claybank dun. The color on the mane, tail, and legs is often slightly darker than the body color. It is not clear how D_- and Cc would act together, but since D dilutes both eumelanin and phaeomelanin, and Cc dilutes only phaeomelanin, it might be expected that D_- would override the effects of Cc. However, with the reddish pigments they should work together.

Flaxen (pale-colored) manes and tails appear to be of two kinds. The most common type seems to be controlled by a pair of recessive genes (*ff*) which are epistatic to genes that control the production of phaeomelanin in the mane and tail. A genotype of a chestnut with flaxen mane and tail would be *CCddeeff* (see Figure 8.3). A chestnut with chestnut mane and tail would be *CCddeeF_*. Black manes and tails would not be affected by *ff*. A second type of flaxen mane and tail, called silver dapple, is apparently controlled by a dominant gene *S* and affects only black. This phenotype is common in Shetland ponies (*aaB_CCddE_S_*). Black horses without silver manes and tails would carry *ss*.

Gray and roan patterns are the result of white hairs interspersed with colored hairs. Horses that are genetically gray have very few white hairs when they are young and become progressively more gray as they mature. Older horses in some breeds may be totally void of the colored hair that they possessed when younger. The whole body, including the head and feet, generally expresses about the same degree of graying. Gray (also called progressive gray) is dominant (*G_*) to nongray (*gg*). Roan horses have about the same proportion of white hairs when they are mature as when they were younger. Roaning is usually less apparent on the head and feet than on the body. Most authorities agree that the roan gene (*R*) is a partially dominant or codominant gene and is an embryonic lethal when homozygous. *Rr* is

Figure 8.3 A Belgian mare with flaxen (straw-colored) mane and tail. (Owner: James Sheehy, Newtonia, MO.)

roan and should not breed true, although there has been at least one report of a true-breeding roan; *rr* is nonroan.

Irregular areas of solid-white spotting imposed over any contrasting color characterize "paint" and "pinto" patterns. Black-and-white patterns are sometimes called piebald, although it may also refer to white spots on any color. "Skewbald" has been used to refer to any nonblack horse with white spots. Several kinds of spotting patterns exist, which can be distinguished one from another by someone with a little experience. The most common kind of spotting, called tobiano, is probably controlled by a dominant gene designated *T.* However, one of the less common types, called overo, is thought to be due to a recessive gene (*o*) at another locus.

8.5.3 Color in Dogs

The majority of common colors and patterns in dogs can be explained by the actions of genes at 10 loci. Genes at five loci have effects similar to those observed in horses. At least four alleles are probably present in the *A* (agouti) series, all of which can be observed in German Shepherds. The top dominant in the series, A^y, produces cream- or gray-colored body with a brown or light brown (sable) saddle. The next allele in the series, a^w (recessive to A^y), produces a pattern very much like the first, but with a black saddle (see Figure 8.4). Some authorities speculate that the coloration of the wolf may be controlled by a gene similar to one of these two alleles. The third allele a^t (recessive to both A^y and a^w), produces black and tan dogs, also known as bicolor. The amount of black varies from a black saddle to black on almost the whole body except the points, as in many Dobermans. The tan color is an intense, reddish tan rather than the light tan or cream observed with A^y_ or a^w_. The bottom allele in the series, *a,* is recessive to the other three and produces solid black when homozygous (*aa*).

Genes at the *B* locus determine whether eumelanin will be black (*aaB_*) or brown (*aabb*), also called chocolate. The gene for brown (*b*) is recessive to *B* and epistatic to *aa*. A dominant gene at the *E* locus allows for the synthesis of eumelanin (black or brown). The recessive allele in the homozygous state (*ee*) causes a change in melanin so that phaeomelanin (red or yellow) is produced. The genotype *ee* is epistatic to *aa*, *B_*, and *bb*. Dogs of the genotype *aaB_E_* will be solid black; *aabbE_* will show solid brown (liver or chocolate), *aaB_ee* will be red or reddish yellow with dark nose and lips; and *aabbee* will be yellow or golden with pink nose and lips.

Diluted colors can be produced by genes at the *C* and *D* loci. A dominant gene at the *C* locus allows for full or intense expression of color, as explained earlier. An allele, called chinchilla (c^{ch}), recessive to *C*, is a phaeomelanin modifier that changes tan to cream, and red and yellow to lighter yellow and golden colors. It has no apparent effect on eumelanin.

Figure 8.4 A dog with a black or dark brown saddle over a light tan or cream-colored body; commonly seen in dogs of German Shepherd breeding.

In this sense, it is similar to the cream dilution gene that produces palomino in horses. The *C* locus in mice and rabbits is called the albino locus, but as in horses, a true albino gene in dogs is very rare or nonexistent. A rare pink-eyed white phenotype has been observed in Pekingese dogs, but it is not known whether it is due to the effects of genes at the albino locus (*cc*), or by genes at another locus.

A recessive dilution gene in the homozygous state (*dd*) acts on both types of melanin. It produces the blue dilution of black sometimes seen in greyhounds and poodles. It also dilutes brown and yellow, the effect on yellow being similar to the chinchilla dilution just described. The effect of the *d* gene is similar to the dun dilution gene in horses, except that in horses it has a dominant effect.

Irregular white spotting, often called piebald spotting, is a recessive trait in dogs (*ss*). The dominant allele *S* results in the absence of spotting. The amount of spotting varies from very little white, usually on the under-side of the body, to the presence of almost complete white. The occurrence of all-white dogs with pigmented eyes apparently results from the effects of a pair of recessive genes at the *W* locus (*ww*). A dominant gene for progressive graying (*G*) is present in dogs, the effects of which are almost the same as described for horses.

Two other genetic effects associated with white are controlled by genes at the T and M loci. A dominant gene T causes the appearance of small pigmented spots or "ticks" on a white background. The ticks are not present at birth, but appear gradually later. This speckled appearance is often observed on Dalmatians and setters. A codominant gene, M, when heterozygous (Mm), increases the amount of white in the coat and produces a blotched or dappled pattern of dilute and intense patches called merle. The homozygote, MM, produces an almost-all-white coat, pale blue eyes, and various defects, including deafness, blindness, and structural abnormalities of the eyes.

8.5.4   Color in Cats

The color of cats in the wild was probably similar to the present-day tabby pattern, consisting of black striping over a yellowish-gray background. This background color is similar to agouti described for rabbits and mice. Agouti is under the control of a dominant gene at the A locus (A_-). The homozygous recessive aa produces black. Mackerel tabby (Figure 8.5), commonly observed in mongrel cat populations, is apparently most like the wild pattern. A second kind of tabby, known as blotched tabby, consists of more irregular striping than mackerel. A third kind of tabby is found in the Abyssinian breed. Striping is almost completely absent except for small amounts on the legs and tail. The three kinds of tabby are controlled by three alleles; T^a for Abyssinian (dominant to T and t), T for mackerel (dominant to t and recessive to T^a), and t for blotched tabby (recessive to both T^a and T).

Figure 8.5   The striped pattern of a mackerel tabby cat. (Owner: Ann Lawrence, Springfield, MO.)

For black (*aa*) to be expressed, a dominant gene at the *B* locus must be present (*B_*). A recessive gene in the homozygous state is responsible for brown or chocolate (*bb*), a situation similar to that in dogs. The recessive gene *b* is epistatic to *aa*. Some authorities suggest the possible existence of a third allele, recessive to *b*, that produces a lighter brown, termed milk chocolate.

Black can also be modified by a gene at another locus, this one on the X chromosome. The production of orange or yellow melanin (phaeomelanin) is caused by the presence of a codominant gene *O* (epistatic to *aa*). The allele, *o*, in the homozygous state in females allows black (*aaB_oo*) to be expressed. Because the genes at the *O* locus are X linked, black males would carry only one gene allowing black to be expressed, *aaB_oY*. The Y represents the Y chromosome, which does not carry an *O* locus. Orange cats are *aaB_OO* (females) or *aaB_OY* (males). The effect of *O* on *bb* is probably similar to the effect on *B_* except that a lighter yellow is probably produced. The effect of *O* seems to be similar to that of the *e* gene in dogs and horses, the differences being that *O* is codominant and X linked. Heterozygous females have yellow or orange hairs interspersed among black or brown hairs similar to the roan effect observed in cattle. This pattern in cats is called tortoiseshell. Since males carry only one X chromosome, they do not normally express the tortoiseshell pattern. The rare occurrence of tortoiseshell males was explained in Chapter Five.

Two mutants at the *C* locus, *c^b* and *c^s*, produce the popular Burmese and Siamese color patterns, respectively. In homozygous Burmese cats (*c^bc^b*), the intensity of black is reduced producing a brownish color. Extremities are darker than the body, and the eyes are sometimes blue. Siamese cats (*c^sc^s*) have an almost-all-white body with dark extremities similar to the himalayan rabbit. These alleles are apparently codominant, so that the heterozygote (*c^bc^s*) produces a phenotype intermediate to the pure Burmese and Siamese patterns. Both alleles are recessive to *C*, which permits full color expression as determined by genes at other loci. As in dogs and horses, the existence of a true albino allele is difficult to establish.

An all-white cat, not an albino, is produced due to the effect of a dominant gene (*W_*), the nonwhite phenotype being *ww*. The eyes may be blue, yellow, or odd-colored. The dominant gene apparently has a pleiotropic effect in that almost half the white cats are deaf. A dominant inhibitor gene (*I*) suppresses the development of melanin such that the hair is white with varying amounts of color in the tips. Phenotypes produced by the effects of this gene are silver (as in the silver tabby), chinchilla, and the "smoke" phenotype of the long-haired Persian.

Irregular white (piebald) spotting seems to occur in cats as a dominant trait (*S_*). In dogs, the occurrence of white spotting appears to be recessive. When white spotting occurs with the tortoiseshell pattern in cats, the orange and black hairs appear as distinct patches of orange and black rather than as an interspersion of colored hairs as described earlier. The resulting three-color phenotype is often referred to as calico. The amount of white varies greatly, from very little to almost all white. This variation in the amount of white expressed in spotted animals seems to be controlled by a number of other genes. This kind of inheritance is discussed in Chapter Ten.

A recessive dilution gene *d*, when homozygous, dilutes black to blue and yellow to cream. A similar gene is found in dogs, rabbits, mice, and rats. Intense, or full color, is produced by *D_*.

STUDY QUESTIONS AND EXERCISES

8.1. Define the terms *epistasis* and *hypostasis*.

8.2. What is the minimum number of pairs of genes needed to describe epistasis? Explain.

8.3. What is meant by the term *classic ratios*?

8.4. Compare the meanings of simple and duplicate epistasis.

8.5. Explain why the type of interaction listed in line 5 of Table 8.1 is not called epistasis?

PROBLEMS

8.1. A dominant epistatic gene *W* in cats results in the production of a white hair coat. The homozygous recessive *ww* allows pigment to be produced. The expression of black and orange hair colors are controlled by a pair of codominant sex-linked alleles. Homozygous females are either orange (*OO*) or black (*oo*); the heterozygote (*Oo*) produces the pattern called tortoiseshell. Male cats can only be black (*o*Y) or orange (OY) since they carry only one X chromosome. **(a)** An orange female was bred to a white male and produced a litter of 3 white females and 2 white males; what is the probable genotype of the white sire of this litter? **(b)** The cats in the F_1 litter were confined and later allowed to breed among themselves, producing 6 white males, 5 white females, 1 orange male, 1 tortoiseshell female, and 1 orange female; what are the genotypes of all kittens in the F_1 generation, and their sire? **(c)** What phenotypes would be expected from matings between the F_1 females and their white sire?

8.2. One explanation for the production of the yellow color of Guernsey cattle is that a recessive gene at the *C* locus, when homozygous (*cc*), dilutes red but does not affect black. Black (*B_*) is dominant to red (*bb*). Angus cattle are black and normally do not carry the dilution gene (*BBCC*). Hereford cattle are *bbCC* while Guernseys are *bbcc*. **(a)** Assume that a group of Angus–Guernsey cross heifers are bred to a purebred Guernsey bull; list the genotypes and phenotypes expected among the calves from this cross. **(b)** Suppose that a large

number of crosses were made among dihybrids, what phenotypes and proportions thereof would be expected among the progeny?

8.3. Two pairs of dominant and recessive genes in hogs (R, r, and S, s) are thought
to control the expression of shades of red. Intense or full expression of red
requires the presence of at least one dominant gene at each of the two loci
($R\_S\_$). The presence of a dominant gene at only one locus (R_ss or $rrS\_$) dilutes
the color to sandy or reddish yellow. The double recessive results in white or
very light cream. **(a)** A sandy-colored sow was bred to a sandy-colored boar
and produced a litter of 1 red, 4 sandy, and 2 white pigs; what are the probable
genotypes of the parents? **(b)** Would it be possible to develop a true-breeding
line of sandy-colored hogs? Explain. **(c)** Suppose that several matings were
made among dihybrids; what proportions of colors would be expected among
the progeny?

8.4. Suppose that a certain cross produced two phenotypes in a 9 : 7 ratio. Calculate
the values of χ^2 and their probabilities of occurrence if this were tested as a
1 : 1 ratio and sample size of **(a)** 16 and **(b)** 32 (refer to Chapter Four.)
(c) Maintaining the 9 : 7 ratio and testing it as a 1 : 1, how large a sample would
be needed to produce a χ^2 value that would be significant at the 5 percent level
of probability?

8.5. Homozygous rose-combed chickens mated to homozygous pea-combed types
produce F_1 progeny having walnut combs. In a certain experiment, a group of
the F_1 walnut-combed hybrids were mated among themselves and produced F_2
progeny that, when combs developed, consisted of 31 with walnut combs, 12
with rose combs, 9 with pea combs, and 4 with single combs. **(a)** What is the
probable phenotypic ratio involved? **(b)** Using appropriate symbols, show the
genotype of the single-combed birds. **(c)** What are the genotypes of the rose-
combed and pea-combed P_1 parents? **(d)** What phenotypic results would be
expected from a dihybrid testcross?

REFERENCES

ADALSTEINSSON, S. 1978. Inheritance of yellow dun and blue dun in the Icelandic
toelter horse. *Journal of Heredity* 69 : 146.

CARVER, E. C. 1984. Coat color genetics of the German shepherd dog. *Journal of
Heredity* 75 : 247.

EVANS, J. W., A. BORTON, H. F. HINTZ, and L. D. VAN VLECK. 1977. *The Horse.*
W.H. Freeman and Company, Publishers, San Francisco. Chapter 15.

HUTT, F. B. 1979. *Genetics for Dog Breeders.* W.H. Freeman and Company, Publishers, San Francisco.

LITTLE, C. C. 1957. *Inheritance of Coat Color in Dogs.* Howell Book House, Inc.,
New York.

OLSON, T. A., and R. L. WILLHAM. 1982. Inheritance of coat coloration and spotting patterns of cattle: a review. *Iowa State University Agriculture and Home
Economics Experiment Station Research Bulletin* 595.

ROBINSON, R. 1977. *Genetics for Cat Breeders*, 2nd ed. Pergamon Press, Inc., Elmsford, N.Y.

SEARLE, A. G. 1968. *Comparative Genetics of Coat Color in Mammals*. Academic Press, Inc., New York.

SNYDER, L. H., and P. R. DAVID, 1957. *Principles of Heredity,* 5th ed. D.C. Heath and Company, Lexington, Mass.

SPONENBERG, D. P., and B. V. BEAVER. 1983. *Horse Color*. Texas A & M University Press, College Station, Tex.

NINE

Linkage
and Chromosome Maps

A typical chromosome may contain hundreds of genes. Any two genes that are a part of the same chromosome are said to be *linked*. All the genes on the same chromosome are part of a common linkage group. Cattle have 30 pairs of chromosomes, so there are 30 linkage groups. The principles and problems presented in Chapters One through Eight involved kinds of inheritance where linkage was not a factor. In this chapter we see how linkage affects normal gene segregation and learn about the effects of linkage on usual phenotypic ratios.

9.1 RECOGNIZING LINKAGE

When dealing with two or more traits affected by genes at two or more loci, it is inevitable that at some point a situation will be encountered where the two loci in question are on the same chromosome. The fruit fly *Drosophila melanogaster,* for example, has only four pairs of chromosomes, and well over 500 genes have been identified. Simple probability suggests that about one of every four two-pair crosses can be expected to involve linkage. The common farm animals have somewhere between 25 and 40 pairs of chromosomes, so the probabilities of encountering linkage situations with two-pair crosses are considerably less. Fewer numbers of specific genes have been identified in larger animals, so the chances of knowingly encountering linkage are further reduced. A number of linkage groups have been identified in rats, mice, rabbits, and chickens. As we learn more about the genetics of these and other domestic species, more genes will be specifically identified as belonging to specific linkage groups. It is important then for breeders of domestic animals to know how to recognize linkage and be able to apply that knowledge to animal improvement.

9.1.1 Expectations When Genes Are Not Linked

To recognize that two genes are linked, results of certain crosses when linkage is not involved will first be reviewed. The dihybrid testcross is commonly used in linkage studies, so let us look at the expected results of that cross when genes are not linked.

The black and red colors in cattle are controlled by a dominant gene *B* and recessive gene *b*, respectively. Another dominant gene, *P*, results in the polled condition, while the recessive allele in the homozygous state (*pp*) produces horns. The dihybrid individual, *BbPp*, mated to the double recessive, *bbpp*, constitutes the dihybrid testcross. Problems of this kind have been discussed and solved several times in previous chapters, so we should be able to predict very quickly that this cross should produce the following results:

$\frac{1}{4}$ *BbPp*   black, polled
$\frac{1}{4}$ *Bbpp*   black, horned
$\frac{1}{4}$ *bbPp*   red, polled
$\frac{1}{4}$ *bbpp*   red, horned

This $1 : 1 : 1 : 1$ ratio of genotypes and, in this case, phenotypes is the standard expectation of the dihybrid testcross when genes are not linked.

Since the two loci are on different chromosomes, the dihybrid cow or bull is expected to produce four kinds of gametes in equal proportions: *BP*, *Bp*, *bP*, and *bp*. A double-recessive animal can produce only one kind of gamete, *bp;* therefore, all calves will receive this gamete. The differences in the phenotypes of the progeny of the dihybrid testcross will depend on which gamete is received from the hybrid parent. The importance of this concept will become more apparent as we get into crosses involving linkage.

9.1.2 Expectations When Genes Are Linked

An excellent way to explain how linkage affects phenotypic ratios is to use an illustration. In mice the gene that allows color to be expressed in the hair coat is a dominant gene represented by the letter *C*. In this example we shall assume that all colored mice are black. The recessive gene in the homozygous state (*cc*) produces the familiar pink-eyed, white albino. On the same chromosome is the locus that carries the recessive gene *f* for frizzy hair coat. The dominant allele *F* results in normal smooth hair.

Suppose that a male from a line of mice with smooth black hair (*CCFF*) is crossed with some females that are homozygous recessive at both loci (*ccff*). This was defined in an earlier chapter as a P₁ cross. Each homozygote will produce only one kind of gamete, so the only kind of offspring that can be produced is the F₁ dihybrid *CcFf*. When the dihybrid mice produce their gametes, since the genes at the two loci are on the same chromosome, there would be a tendency to expect genes *C* and *F* to segregate together and genes *c* and *f* to segregate together, as in the following diagram:

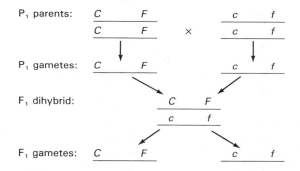

Based on what has been presented so far, we would not expect to see any gametes having both a dominant and a recessive gene. It would seem that $\frac{1}{2}$ the gametes should contain the two dominant genes *C/F,* and the other $\frac{1}{2}$ should contain the two recessive genes *c/f,* with no *C/f* or *c/F* gametes being produced. If this were true, the testcross progeny with these genes linked would consist of

$\frac{1}{2}$ *CCFF*   black, normal hair
$\frac{1}{2}$ *ccff*   albino, frizzy hair

Notice that these are the same two combinations of phenotypes that appeared in the P₁ parents. For this reason, these two combinations are referred to as the *parental phenotypes.* What has been presented in this paragraph is hypothetical because the testcross progeny is not limited to just the two parental phenotypes. Some nonparental pnenotypes are produced by the phenomenon called crossing-over.

9.1.3 Crossing-over

This is an event that occurs between nonsister chromatids of homologous chromosomes while they are in synapsis during late prophase and metaphase of meiosis. It involves an actual exchange of chromosome material.

The principle is illustrated in the following diagram involving the F_1 dihybrid:

Chromosomes before crossing-over	Crossing-over	Chromosomes after crossing-over

Each of the chromatids will ultimately find their way into a separate gamete, so gametes of all four types will be expected to exist. The two carrying the genes C/F and c/f will be called *parental gametes* since they are identical to those produced by the P_1 parents. The other two gametes, C/f and c/F, were produced by crossing-over, so are referred to as crossover gametes. They may also be referred to as *recombinant gametes*. When these two gametes are fertilized by the double-recessive gamete from the testcross parent, they produce the two recombinant phenotypes. The recombinant phenotypes are the two that are different from the parental phenotypes.

When the experiment was actually performed, the testcross results were similar to the following:

53 *CcFf*	black, normal	} 79.4% parental phenotypes
51 *ccff*	albino, frizzy	
12 *Ccff*	black, frizzy	} 20.6% recombinant phenotypes
15 *ccFf*	albino, normal	
131 total progeny		

The recombinant phenotypes were produced by crossing-over. The percentage of recombination, in this example 20.6 percent, is also referred to as the percentage of crossing-over. It is merely the number of recombinant phenotypes expressed as a percentage of the total number of progeny.

9.1.4 Proof of Linkage

If these genes had not been linked, the ratio among the testcross progeny would have been 1 : 1 : 1 : 1, and the percentage of parental and recombinant phenotypes would have been 50 percent each. Evidence indicating that linkage is involved lies in the fact that the ratio deviates significantly from the classic 1 : 1 : 1 : 1, or more specifically, it deviates from a ratio of 50 percent parentals to 50 percent recombinants. Note that when this deviation occurs due to linkage, it is always in the direction of greater numbers of parentals and fewer numbers of recombinants.

The maximum percentage of recombination expected without linkage is 50 percent; therefore, any testcross ratios having fewer than 50 percent recombinants would suggest that linkage is involved. The question then arises as to how much deviation from the 50 : 50 ratio of parentals to recombinants can be expected due to chance when genes are not linked. The chi-square test can be administered to determine this. The actual numbers, not percentages, of testcross progeny would be compared to numbers expected had the ratio been a perfect 1 : 1 : 1 : 1. In the problem used in the illustration above, the deviations from the classic ratio were great enough that the test was not necessary. Most people would conclude intuitively that it would be highly improbable for deviations of this magnitude to occur due to chance. The chi-square test would show deviations of this order to occur due strictly to chance less than once in 10,000 such experiments.

9.2 CHROMOSOME MAPS

Linkage studies not only help determine which genes may or may not be in certain linkage groups, they can also help determine the relative positions of the various loci on the chromosomes. The latter principle is called *chromosome mapping.* Rather extensive chromosome maps have been drawn for humans, rats, rabbits, mice, chickens, and particularly for the fruit fly. Extensive maps have also been prepared for several domestic plants, including corn.

9.2.1 Map Distance

The distance between two loci on a chromosome is proportional to the amount of crossing-over that can be detected in a properly conducted testcross. Stating it another way, the percentage of crossing-over that occurs is proportional to the distance between the genes on the chromosomes.

With but a few exceptions, crossing-over occurs with equal frequency between nonsister chromatids during synapsis at any point on the chromosome. At least three exceptions are noted. Crossing-over is less likely to occur near the centromere, near the ends of the chromosomes, and near another crossover. For the most part, if crossing-over occurs anywhere along the chromosomes with approximately equal frequency, crossover gametes will be produced only when the crossover site occurs between the two loci in question. If two gene loci are located relatively close together on the chromosomes, crossing-over will be less likely to occur between them, so recombinant gametes are less likely to be produced. When linked genes are located relatively far apart on the chromosomes, a greater number of cross-over gametes will be expected. This is a very important principle regarding linkage; it might be well to reread this paragraph to be sure that it is understood.

A testcross producing 10 percent recombinants indicates that the genes are linked and located closer together on the chromosomes than in another example of a testcross producing 20 percent recombinants. If it is known that two gene loci are located on the ends of the same chromosome, every crossover that takes place would occur between the two loci and produce crossover gametes. Since a single crossover would occur between one of the two chromatids on one chromosome and one of the two chromatids on the homolog, the maximum number of crossover gametes that could be produced would be one-half the total number of gametes. In other words, if the genes were located on the ends of the chromosomes, there would be one parental gamete produced for every crossover gamete. It would be as if the genes were actually located on different chromosomes. The closer the genes were on the chromosomes, the fewer the crossover gametes that would occur and the greater the number of parental gametes that would occur. The theoretical range of crossover percentage that can exist with two-point test-crosses is zero to 50, while the theoretical range of parentals is 50 to 100 percent.

Since crossover percentages are proportional to distances between linked genes, these figures can be used as a quantitative measure of map distance. Percent crossing-over (or percent recombination) is equivalent to map units. Since the maximum percentage of crossing-over that can occur is 50 percent, the number 50 can be chosen as the arbitrary length of any chromosome. The actual length of the chromosome in angstrom units or nanometers is not important. One map unit is defined as $\frac{1}{50}$ the length of any chromosome. The relative length of a chromosome is 50 map units. In the example presented earlier, percent crossing-over was calculated to be 20.6; the distance between the C and F loci is therefore estimated to be 20.6 units.

9.2.2 Mapping of Chromosomes

The knowledge that two genes are located 20.6 units apart on a pair of homologous chromosomes that are 50 units long does not identify their actual locations. More information must be obtained concerning the relative distances between each of these two genes and other genes known to be part of the same linkage group.

The example presented in the next few paragraphs will illustrate how a knowledge of the distances among several genes will help in identifying their probable locations on the chromosomes. In mice the recessive gene d

dilutes black to blue-gray. The dominant allele *D* allows black to be expressed. Another recessive gene, *w,* produces curly whiskers, while the dominant allele *W* results in normal straight whiskers. A cross between homozygous blue-gray mice with straight whiskers (*ddWW*) and homozygous black mice with curly whiskers (*DDww*) produced a litter of black mice with straight whiskers. The F₁ dihybrids (*DdWw*) were later mated to blue-gray mice with curly whiskers (testcross). Assume that the testcross resulted in the following phenotypes and numbers among the progeny:

15	blue-gray mice with straight whiskers
13	black mice with curly whiskers
8	black mice with straight whiskers
9	blue-gray mice with curly whiskers
45	total progeny

Since the ratio is clearly not a 1 : 1 : 1 : 1, linkage is indicated. The distance between the two loci can be estimated by calculating the percentage of crossing-over. Looking back at the original homozygous parents (P₁), it can be seen that the parental phenotype is blue-gray with straight whiskers and black with curly whiskers. The other two combinations of phenotypes are the recombinants. Notice that in this example, the dihybrid and the double-recessive genotypes are associated with the recombinant phenotypes. This will not always be the case. It must be remembered, however, that the test for linkage is to determine how the genes are linked in the homozygous P₁ generation and not in the testcross individuals. The percentage of recombinants is 37.8. This is determined by counting the number of recombinant phenotypes and expressing it as a percentage of the total number of progeny produced, 45 in this case.

We know now that the genes at the *D* and *W* loci are located on the same chromosome approximately 38 units apart. Since the total length of the chromosomes is assumed to be 50 units, we do not know their actual locations, only the distance between them. We also know that in the parents of the dihybrid, the dominant gene *D* is linked with the recessive gene *w* in one parent, and the recessive gene *d* is linked with the dominant gene *W* in the other. This is apparent because their genotypes are *DDww* and *ddWW;* there is no other way for them to be linked.

To obtain more information as to the actual locations of these genes, we can conduct testcrosses involving other genes known to be on the same chromosomes. Another locus, which will be called the *K*-locus, is known to be in the same linkage group. The recessive gene in the homozygous state (*kk*) causes a mouse to have a distinctive kink in its tail. The dominant allele *K* results in the normal straight tail. Assume that the P₁ parents that were black also had kinked tails, while the blue-gray parents had straight tails. Considering only the genes at the *K* and *D* loci, suppose that a testcross of the dihybrid produced the following results:

35 blue-gray mice with normal tails
33 black mice with kinked tails
1 blue-gray mouse with kinked tail
$\underline{2}$ black mice with normal tails $\Bigg\}$ (4.2% crossing-over)
71 total progeny

The approximately 4 percent crossing-over indicates that the loci are located relatively close together on the chromosomes. This example also illustrates the importance of the size of the sample of testcross progeny. If the size of the sample had been smaller, it is quite possible that due to chance, no crossover gametes might have been produced. If, for example, the genes were linked 1 unit apart, a sample of 100 would be needed to expect only one recombinant to be produced.

We know now that in one of the P_1 parents the gene D is located about 38 units from w, and about 4 units from k. In the other P_1 parent the same distances are involved, but with the genes d, W, and K. We do not know the distance between the W and K loci. Logic would indicate that it is either 34 units, the difference between the other known distances, or 42 units, the sum of the other distances. Let us assume that another testcross involving the K and W loci produced the following results:

14 mice with normal tails and normal whiskers
14 mice with kinked tails and curly whiskers
11 mice with kinked tails and normal whiskers
$\underline{9}$ mice with normal tails and curly whiskers
48 total progeny

The 20 recombinants represent 41.7 percent of the total progeny and translate into approximately 42 map units. Our earlier prediction was that it would be either 34 or 42 units. It can be concluded now that the D locus is between the K and W loci. A chromosome map showing the positions of these loci in the dihybrid might look something like this:

W	38	d 4 K
w		D $\quad$ k

Notice that the exact position of each locus is still not identified. The W locus could be located anywhere from the end of the chromosome to about 6 units from the end. Additional testing involving other genes located on the same chromosome should be conducted to resolve this question.

9.2.3 Coupling versus Repulsion

When genes in the P_1 parents are linked dominant to dominant and recessive to recessive, they are said to be linked in the coupling phase. Suppose that a P_1 cross of *AABB* × *aabb* is made where the genes are on the same pair of chromosomes 24 units apart. A testcross of the F_1 dihybrid would produce the following ratio of genotypes and phenotypes:

$$\left.\begin{array}{l} 38\%\ AaBb \\ 38\%\ aabb \end{array}\right\}\ 76\%\ \text{parental phenotypes}$$

$$\left.\begin{array}{l} 12\%\ Aabb \\ 12\%\ aaBb \end{array}\right\}\ 24\%\ \text{recombinant phenotypes}$$

If the genotypes of the P_1 parents had been *AAbb* and *aaBB,* the F_1 testcross would have produced the following combination:

$$\left.\begin{array}{l} 12\%\ AaBb \\ 12\%\ aabb \end{array}\right\}\ 24\%\ \text{recombinant phenotypes}$$

$$\left.\begin{array}{l} 38\%\ Aabb \\ 38\%\ aaBb \end{array}\right\}\ 76\%\ \text{parental phenotypes}$$

In the second case, the genes were linked in the repulsion phase, dominant with recessive, and recessive with dominant. Notice that in both examples, the parental group of phenotypes is larger than the recombinant group. In cases where the phenotypes of the P_1 generation are not known, and a testcross indicates the presence of linkage, the parental group of phenotypes can be identified as the larger of the two groups.

9.2.4 Multiple Crossing-over

It is possible for more than one crossover to occur between the same two chromatids of the same chromosome. It is not likely to occur within approximately 5 to 10 map units of another crossover, however. If two crossovers were to occur between two genes on the same two chromatids, one crossover would tend to cancel the other, producing a parental gamete. The greater the distance between any two linked genes, the greater the chances that a second or third crossover will occur between them. Odd numbers of two-strand crossovers produce recombinant gametes; even numbers of crossovers tend to restore the parental combinations. This is illustrated in the following diagram of a double crossover:

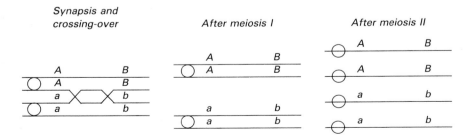

Note that although there is linkage and crossing-over, there are no crossover or recombinant gametes. The occurrence of two crossovers has restored the parental combinations. The implication is that when the percentage of recombination is calculated in linkage tests, none of the double crossovers will be taken into account. This will result in an underestimation of map distance between any two loci. For map distances below about 20, the errors will be relatively small. The greater the calculated map distance above 20, the greater will be the error.

To detect a double crossover, a third locus situated between the other two must be identified. The following diagram shows how double crossovers can be detected by using a third marker:

Notice that double crossing-over has shifted the location of the genes at the D locus from one chromatid to another, resulting in crossover gametes which would, in turn, produce detectable recombinant phenotypes in testcross progeny.

Double crossovers between the A and D loci or between the B and D loci would not be as likely to occur because they are closer together; therefore, two-point testcrosses should result in relatively accurate measures of the map distance between them. A two-point testcross involving the A and B loci would more likely involve double crossovers and result in a smaller calculated percentage of recombination. For example, if it were determined from two-point testcrosses that the distance between A and D were 15 units and the distance between D and B were 20 units, logic would suggest that the percent recombination in a testcross involving A and B should be about 35. However, such a testcross would more likely result in a number closer to 30. On the other hand, although the longer distances might be underestimated, the prediction of the linear order of the genes would not be affected.

9.2.5 Three-Point Testcross

Testcrosses involving genes at three different loci can be used to determine linkage patterns and map distances. They have the advantage also of revealing the occurrence of double crossovers. To illustrate, let us use data from a three-point testcross in rabbits conducted by W. E. Castle, who contributed so much to the present knowledge of animal genetics.

Three loci, arbitrarily identified as the *B, C,* and *Y* loci, are involved carrying three pairs of dominant and recessive genes. Black hair coat (*B_*) is dominant to brown (*bb*); the expression of full color (*C_*), in this case black or brown, is dominant to the himalayan pattern (*cc*); the presence of white fat (*Y_*) is dominant to yellow fat (*yy*). The testcross, *BbCcYy* × *bbccyy,* produced 908 rabbits, including the following phenotypes:

> 276 black, himalayan, white fat
> 7 black, himalayan, yellow fat
> 55 black, full color, white fat
> 108 black, full color, yellow fat
> 125 brown, himalayan, white fat
> 46 brown, himalayan, yellow fat
> 16 brown, full color, white fat
> 275 brown, full color, yellow fat

If the genes at the three loci were not linked, the expected numbers of each phenotype should have been closer to 113 or 114. Since they are not, it is assumed that they are linked. Recall that crossover combinations are less likely to occur than parentals, and double crossovers are less likely to occur than single crossovers. It can be concluded that the phenotypes associated with the numbers 275 and 276 are parentals, and those associated with the 7 and 16 are the double crossovers. The other combinations are due to single crossovers.

Looking at the crossover gametes produced by the trihybrid will help in identifying the linkage pattern:

Parental types	Single crossovers	Single crossovers	Double crossovers
276 *BcY*	108 *BCy*	55 *BCY*	7 *Bcy*
275 *bCy*	125 *bcY*	46 *bcy*	16 *bCY*
551 60.7%	233 25.7%	101 11.1%	23 2.5%

Comparing the double-crossover gametes to the parentals will help determine the linear order of genes on the chromosomes. The gene in the double-crossover gametes which has been transposed compared to the parental gametes is the one that is in the middle. Making this comparison shows that it is the genes at the *Y* locus that were transposed, indicating that the *Y* locus is between the other two. Thus the order of the loci is *B-Y-C.*

The particular linkage pattern in the P_1 parents and the trihybrid can be determined by noting the traits expressed in the parental phenotypes. In one parental type, black, himalayan, and white fat are associated. The gametes from the trihybrid individuals that would produce this combination would contain the genes *BYc.* The other parental type—brown, full color, and yellow fat—would have been produced by the parental gamete *byC*. The genotypes of the P_1 generation would have been *BBYYcc* and *bbyyCC*.

The single crossover percentage of 25.7 occurred in the region between genes *B* and *Y* and *b* and *y*. However, the distance between these two loci is greater than 25.7 map units because some additional crossovers have been counted in the double-crossover percentage. The total percentage of crossing-over between loci *B* and *Y* is actually the sum of 25.7 and 2.5. The map distance in this region, then, is 28.2 units. Similarly, the distance between the *C* and *Y* loci is 11.1 plus 2.5, or 13.6 map units. The calculated distance between the outside loci is simply the sum of the other two distances, 41.8 units. If a two-point testcross had been conducted involving genes at the *B* and *C* loci, the estimation of the distance between them would have been closer to 36 or 37.

A diagram of the linkage pattern and chromosomes as they would be in the trihybrid might look like this:

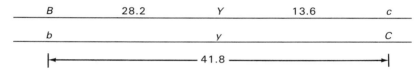

9.3 OTHER FACTORS ASSOCIATED WITH LINKAGE

Since no principle of genetics functions completely apart from other genetic principles, it is important to know what to expect when any two or more principles are working together. In the next few paragraphs we examine the effects of linkage on sex-linked genes, dihybrid ratios, and epistasis, and try to distinguish between very closely linked genes and plieotropy.

9.3.1 Linkage and Sex-Linked Genes

Genes located on sex chromosomes are members of a common linkage group just as are those on any of the autosomes. The gene for barred feathers (*B*) in chickens is located on the Z chromosome and is dominant to the allele for nonbarred feathers (*b*). A dark streak that appears in the down on the head of chicks of some breeds is controlled by a Z-linked recessive gene *h*. The dominant allele *H* results in the absence of the head streak. It is known that the loci for these genes are 13 units apart on the chromosome. Assume that males of the genotype *BBHH* were crossed with females of the genotype *bWhW*. The F_1 birds would be composed of approximately equal numbers of normal barred dihybrid males (*AaHh*) and normal barred females (*BWHW*). A testcross would involve crossing the dihybrid males to a number of hens of the genotype *bWhW*. The testcross progeny would consist of the following eight genotypes in the specified percentages:

	Males	Females	Phenotypes	
43.5%	*BbHh*	*BWHW*	barred, no head streaks	
43.5%	*bbhh*	*bWhW*	nonbarred, streaked heads	} 87% parentals
6.5%	*Bbhh*	*BWhW*	barred, streaked heads	
6.5%	*bbHh*	*bWHW*	nonbarred, no head streaks	} 13% recombinants

Compared to autosomal linkage studies, the major difference is that the dihybrid does not exist in both sexes. In birds, it is a male; in mammals it is a female. The test mate to which the dihybrid is bred in each case is the double-recessive individual.

9.3.2 Effect of Linkage on F_2 Ratios of Dihybrid Crosses

To illustrate the effect of linkage on the usual 9 : 3 : 3 : 1 expected from a dihybrid cross with two sets of dominant and recessive genes, an example will be taken from laboratory rats. A recessive mutant gene *k* produces kinky hair when in the homozygous state. The dominant allele *K* results in the normal smooth hair. Another recessive mutant gene *s* produces a short stubby tail when homozygous. The dominant allele *S* produces a tail of normal length. The genes are linked approximately 34 units apart. The percent recombination from a testcross would be expected to be 34, the same as the map distance.

Assume that in two pure lines of rats, the genes are linked in the repulsion phase, *KKss* and *kkSS*. The dihybrids from a cross of these two lines would be expected to produce gametes in a ratio of 66 percent parental types (about $\frac{2}{3}$) and 34 percent recombinant types (about $\frac{1}{3}$). Four types of gametes would be expected: *K/S, K/s, k/S,* and *k/s,* but in a ratio of about 1 : 2 : 2 : 1, respectively, rather than 1 : 1 : 1 : 1, the expected ratio without linkage. The proportion of the 16 possible combinations from the dihybrid cross can be illustrated using a Punnett square:

	1 *K/S*	2 *K/s*	2 *k/S*	1 *k/s*
1 *K/S*	1 *KKSS*	2 *KKSs*	2 *KkSS*	1 *KkSs*
2 *K/s*	2 *KKSs*	4 *KKss*	4 *KkSs*	2 *Kkss*
2 *k/S*	2 *KkSS*	4 *KkSs*	4 *kkSS*	2 *kkSs*
1 *k/s*	1 *KkSs*	2 *Kkss*	2 *kkSs*	1 *kkss*

The proportion of each of the four phenotypic classes would be as follows:

```
19 K_S_ or 52.78% with normal hair and normal tails   (recombinant)
 8 K_Ss or 22.22% with kinky hair and normal tails  ⎫
 8 kkS_ or 22.22% with normal hair and stubby tails ⎬ parentals
 1 kkss  or  2.78% with kinky hair and stubby tails   (recombinant)
36 total    100.00%
```

The total percentage of recombinants is 55.66, compared to 62.5 expected without linkage. Note that the percentage of recombinants is less with linkage than without linkage, just as it was with the testcross data.

9.3.3 Effect of Linkage on Epistatic Ratios

In Chapter Eight it was shown how classic ratios can be modified by epistasis. In this chapter we have seen how those ratios, particularly dihybrid testcross ratios, can be modified by linkage. Let us now look at an example of how both effects can work together and how their effects can be distinguished.

In rabbits, black hair color ($B-$) is dominant to brown color (bb). Albinism is a homozygous recessive trait (cc), while the dominant allele ($C-$) allows color to be produced. Assume that a homozygous black rabbit ($CCBB$) is mated to an albino of the genotype $ccbb$. The F_1 progeny are all dihybrids ($CcBb$). To determine whether or not the genes are linked, the usual testcross would be performed, $CcBb \times ccbb$. If the genes are not linked, the following results would be expected:

$$\frac{1}{4} \, CcBb \qquad \frac{1}{4} \text{ black}$$
$$\frac{1}{4} \, Ccbb \qquad \frac{1}{4} \text{ brown}$$
$$\left.\begin{array}{l}\frac{1}{4} \, ccBb \\[4pt] \frac{1}{4} \, ccbb\end{array}\right\} \quad \frac{1}{2} \text{ albino}$$

Because of epistasis, the classic testcross ratio of $1:1:1:1$ is not expected; instead, the expected ratio without linkage is a $1:1:2$.

When the test cross was actually conducted, the results were similar to this:

```
 62  brown
 87  black
147  albino
296  total
```

It is apparent that epistasis is at work because two of the phenotypic classes have been combined. However, if epistasis were the only factor involved, the ratio would have been closer to $1:1:2$. The actual results show a greater number of black rabbits, a parental phenotype, than brown, a recombinant phenotype, an indication of linkage.

The albino phenotype cannot properly be classified as a parental phenotype since epistasis has modified the expression of the color genes. To determine the percent crossing-over, only the black and brown rabbits can be used in the calculations. Among the 149 rabbits that are not albinos, 62 are recombinants. Crossing-over is calculated as 41.6 percent and the distance between the two loci on the chromosomes is 41.6 map units.

9.3.4 Comparing Linkage to Pleiotropy

Pleiotropy refers to the effect of one gene on two or more traits. An example of such a gene is one in mice that is lethal when homozygous but produces yellow hair color when heterozygous. Yellow mice are usually more obese than their nonyellow litter mates, due to a lower metabolic rate. They have larger skeletons and are less susceptible to certain kinds of cancer. These effects are apparently all due to the presence of one gene.

If two genes are located so close together on the same chromosome that crossing-over is unlikely to occur between them, the effects of the two genes could be mistaken for the pleiotropic effect of a single gene. The locus that carries the gene for rose comb in chickens is only 0.4 map unit from the creeper gene locus. The rose-comb trait (R_-) is dominant to single comb (*rr*). The creeper condition, which was described in Chapter Two, results from the heterozygous state of two codominant alleles (*Cc*). One homozygote (*cc*) produces the normal condition, while the other homozygote (*CC*) results in death of the embryo in very early development.

Assume that dihybrids in which the genes were linked R/C and r/c were mated to birds of the genotype *rrcc*. If the genes were not linked, equal numbers of the following four phenotypes would be expected:

rose-combed creepers } single-combed normals }	50% parentals
rose-combed normals } single-combed creepers }	50% recombinants

However, since the genes are linked, fewer numbers of recombinants would be expected. In this instance, given the closeness of linkage, the percent crossing-over expected would be only 0.4. To expect one each of the two recombinant phenotypes in the cross above, approximately 500 testcross progeny would be needed. Even a sample of 100 progeny might not produce any of the recombinant phenotypes. As long as the creeper trait were to be expressed with the rose-comb trait, the conclusion might be made that they were under the control of the same gene. The greater the number of progeny produced by the dihybrids, the greater the probability that crossing-over will ultimately occur between the two loci, revealing that it was indeed a case of close linkage rather than an example of pleiotropy.

STUDY QUESTIONS AND EXERCISES

9.1. Define each term.

coupling	map unit	recombinant
crossing-over	parental	repulsion
linkage	pleiotropy	testcross
linkage group		

9.2. Explain why linkage should play a more important role in fruit flies than in cattle.

9.3. How would you recognize that genes are linked after conducting a dihybrid testcross?

9.4. If two genes are indeed linked, it would seem logical to assume that only parental gametes would be produced by a dihybrid, but we know that this is not the case. Explain the source of the nonparental gametes.

9.5. Suppose that the results of a testcross are not a perfect 1 : 1 : 1 : 1, what can be done to help decide whether or not it is a theoretical 1 : 1 : 1 : 1?

9.6. What is the relationship between map distance and the incidence of crossing-over? Explain.

9.7. Why is the length of a chromosome assumed to be 50 units rather than 100 or some other number?

9.8. Discuss at least two problems associated with linkage studies of genes that are located very close together on the chromosomes.

9.9. In the procedure for testing for linkage, why are testcrosses used instead of dihybrid crosses?

9.10. How do effects of linkage and epistasis on testcross results differ?

PROBLEMS

9.1. Black in hogs is controlled by a dominant gene, while red is recessive. The gene for erect ears is dominant to its allele for droopy ears. Hampshires are assumed to be homozygous for both dominant genes, while Durocs are homozygous for both pairs of recessive genes. When some F_1 dihybrid Duroc-Hampshire sows were bred to purebred Duroc boars, the litters were composed of the following phenotypes and numbers:

> 17 black pigs with erect ears
> 19 black pigs with drooping ears
> 16 red pigs with erect ears
> 20 red pigs with drooping ears
> ――
> 72 total pigs

(a) Are the genes linked? **(b)** If it were assumed that the genes are linked, what is the distance between them?

9.2. Himalayan is a color pattern in mice and rabbits where the body is mostly white with colored extremities. In mice this trait is recessive to the expression of full color, which in this problem will be black. Use C as the symbol for the dominant gene and c for the recessive gene. The gene for frizzy hair coat (f) in mice is recessive to normal hair. Assume that several dihybrid mice were testcrossed and produced the following progeny:

> 12 frizzy black mice
> 14 smooth himalayans
> 4 frizzy himalayans
> 3 smooth black mice

(a) Are these genes linked? (b) If the genes are linked, what is the percent crossing-over? (c) What were the genotypes of the parents of the dihybrid? (d) Test the observed ratio against a 1 : 1 : 1 : 1 ratio using a chi-square analysis (see Chapter Four); what is the probability that at least this much deviation would be expected due to chance?

9.3. In chickens, three loci, which will be identified as B, C, and F, are on the same pair of homologous chromosomes. The dominant gene B produces white feathers and the recessive allele b produces black when homozygous. A dominant gene C results in a crested bird, while the homozygous recessive cc is noncrested. A dominant gene F produces normal smooth feathers, while the recessive homozygous ff produces frizzled feathers. Three separate two-point test-crosses produced these results:

16 frizzled white	11 frizzled noncrested	30 white noncrested
17 normal black	13 normal crested	34 black crested
4 frizzled black	5 frizzled crested	3 black noncrested
3 normal white	5 normal noncrested	5 white crested
40 total	34 total	72 total

(a) Construct a partial map of the chromosomes in the dihybrid. (b) What were the genotypes of the original P_1 generation?

9.4. In humans, a dominant gene E causes oval-shaped red blood cells, a condition known as eliptoerythrocytosis. The normal-shaped red cells are the result of the homozygous recessive ee. The presence of Rh antigens on the blood cells is due to the effect of a dominant gene R and is called Rh positive. Rh-negative people are homozygous recessive, rr, and have no antigens on the red cells. It is known that the genes at the Rh locus are linked with those at the E locus about 20 units apart. Suppose that a man known to be heterozygous at both loci is married to a woman who has normal red cells and is Rh negative. The man's mother was Rh negative and had normal red blood cells. (a) What probable combination of phenotypes constitutes the parentals? (b) If this couple were to have a daughter, what is the probability that she would be Rh negative and eliptocytotic? (c) If this couple were to have a large family, what proportion of the children are likely to be eliptocytotic? (d) Suppose that due to chance, all of their children have normal red blood cells; what proportion of them are likely to be Rh negative?

9.5. A recessive sex-linked gene f causes an increase in feathering rate in very young chicks. The dominant allele F results in the usual rate of feathering. Another recessive, s, results in gold plumage, the dominant allele produces silver plumage. A cock with silver plumage and normal feathering rate whose dam was gold and expressed fast feathering was mated to several fast feathering hens with gold plumage. The results among the male progeny of these crosses were:

> 23 fast-feathering silver chicks
> 3 fast-feathering gold chicks
> 4 slow-feathering silver chicks
> 20 slow-feathering gold chicks

(a) Based on the information presented, are the genes for gold and silver plumage sex-linked? (b) What is the distance between the F and S loci?

REFERENCES

ALTMAN, P. L., and D. S. DITTMAR, eds. 1972. *Biology Data Book,* Vol. 1. Federation of the American Society of Experimental Biology, Bethesda, Md. (Linkage groups for vertebrates, pp. 15–58)

CASTLE, W. E. 1936. Further data on linkage in rabbits. *Proceedings of the National Academy of Sciences* 22 : 222. Cited by Hutt and Rasmussen, 1982.

GREEN, E. L., ed. 1966. *Biology of the Laboratory Mouse.* Dover Publications, Inc., New York.

HUTT, F. B. 1949. *Genetics of the Fowl.* McGraw-Hill Book Company, New York.

HUTT, F. B., and B. A. RASMUSSEN. 1982. *Animal Genetics,* 2nd ed. John Wiley & Sons, Inc., New York. Chapter 10.

KING, R. C., ed. 1975. *Handbook of Genetics,* Vol. 4, *Vertebrates of Genetic Interest.* Plenum Press, New York.

STANSFIELD, W. D. 1983. *Theory and Problems of Genetics,* 2nd ed. McGraw-Hill Book Company, New York. Chapter 6.

PART 3
QUANTITATIVE INHERITANCE

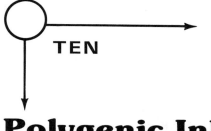

TEN

Polygenic Inheritance

Polygenic inheritance literally refers to many genes affecting a single trait. Multiple genes, not to be confused with multiple alleles, is another term sometimes used to mean the same thing. Since many pairs of genes are involved, those at any one locus will generally have a very small effect. Most of the traits in domestic animals that are considered economically important are affected by this kind of inheritance. Much more will be said about it in this and succeeding chapters.

10.1 QUALITATIVE AND QUANTITATIVE INHERITANCE

All the examples of traits that have been discussed in the previous chapters have been the kind where phenotypes were rather easily distinguished one from another. Cows were either black or red or white; they were spotted or they were not; they had horns or they did not. Hogs had a white belt or they did not; they had droopy ears or their ears were erect. The distinctions were easy to make without the use of special measuring tools. These kinds of traits are called *qualitative traits* and are controlled by only one or a few pairs of genes. The interaction between the genes involved varying degrees of dominance and epistasis.

Another kind of inheritance that exists in domestic animals is called *quantitative inheritance.* The differences among separate phenotypes for a quantitative trait may be so small that they cannot be distinguished by casual observation. The difference between a steer gaining 2.9 pounds (lb) per day and one gaining 3.0 lb is very real, but in order to detect and distinguish between them, the animals must be weighed several times and certain calculations made. The differences in the amounts of milk produced by dairy cows must be weighed to evaluate and compare these cows accurately.

For quantitative traits, the variation observed among large groups of animals is described as *continuous*. For example, Thoroughbred racing horses are not merely identified as fast horses or slow horses; their speeds are measured in increments of fractions of seconds. Within certain normal limits, among a large number of horses, literally any rate of speed can be observed. Seldom will any two horses be timed at exactly the same rate. If all horses were either fast or slow, with no variable rates of speed between, no clock would be needed to distinguish between them. This would then be a qualitative trait expressing discontinuous variation.

Quantitative traits are controlled by many pairs of genes, each having only a small effect. The relationship between or among alleles affecting quantitative traits is usually one of codominance or lack of dominance. Some genes would add to, or make positive contributions to the trait; others would not. Those that would increase the magnitude of value of the trait are sometimes called *additive genes*. For this reason, this kind of gene action is often called *additive gene action*. This is in contrast to dominance and epistasis, which are called *nonadditive gene action*.

It should be kept in mind that polygenes segregate independently and have specific effects, just as do genes with dominant and epistatic effects, but the magnitude of the effect on the phenotype may be quite different. One purpose of this chapter is to help the reader make the transition from thinking in terms of one or two pairs of genes having large qualitative effects to thinking of many genes, each having small quantitative effects. Examples in the next few paragraphs have been selected to help carry out that process.

10.2 NATURE AND DESCRIPTION OF POLYGENIC INHERITANCE

10.2.1 An Example from Cattle

Most examples of polygenic inheritance involve many pairs of alleles residing at as many loci; however, the concept can be illustrated simply using two pairs of alleles and two loci. The amount of white on spotted cattle such as Holsteins will be used to illustrate. The variation in spotting in this breed is such that some cattle have so little white that they are almost all black, while others are almost all white. Assume that two pairs of alleles are involved, A and a, and E and e. Let the genes A and E be those that add to the amount of spotting or contribute more white. The alleles a and e will be considered as neutral genes that add nothing. Further assume that there is a total lack of dominance between alleles, and each additive gene adds the same amount of white. Each additive gene (A and E) will be prescribed as increasing white by an amount equal to 20 percent of the animal's surface. Table 10.1 lists the possible genotypes and degrees of spotting possible based on these assumptions. The genotype $aaee$ will produce the minimum amount of white; the genotype $AAEE$ will produce the most.

If we assume that the genes are present in the population in about equal frequencies, there should be more cattle with 50 percent white and 50 percent black than any other combination, partly because there are more genotypes for that phenotype. One way to get an idea of how many of each phenotype would be expected in a random breeding population is to look at the results of the dihybrid cross. Crossing cattle that are almost all black, $aaee$, with those that are almost all white, $AAEE$, should produce dihybrid genotypes, $AaEe$. The genotypic results would be the same as for all dihybrid crosses:

1 *AAEE*	2 *AaEE*	1 *aaEE*
2 *AAEe*	4 *AaEe*	2 *aaEe*
1 AAee	2 Aaee	1 aaee

4   +   8   +   4   =   16 total

The number of additive genes, whether A or E, will determine the degree of spotting. Table 10.1 lists the phenotypic results of the dihybrid cross.

In reality, it is not known exactly how many loci are involved, and whether each additive gene actually contributes the same as any other. It is possible that small degrees of dominance may exist between alleles, and more than two alleles may be involved at some loci. The effects of genes might well be modified by linkage, epistasis, or environment. Imposing cer-

TABLE 10.1   Inheritance of Degrees of Spotting in Cattle (Hypothetical)

Number of Additive Genes	Possible Genotypes	Degree of Spotting	Results of Dihybrid Cross
0	*aaee*	10% white, 90% black	1
1	*Aaee, aaEe*	30% white, 70% black	4
2	*AAee, aaEE, AaEe*	50% white, 50% black	6
3	*AAEe, AaEE*	70% white, 30% black	4
4	*AAEE*	90% white, 10% black	1
			‾‾
			16

tain limiting assumptions makes the principle easier to illustrate. Once the principle is understood, it must be realized that some of the assumptions may only be partially accurate. The implication is that the distinctions between any two phenotypes may not be as clear cut as was suggested in the example.

10.2.2 Gene Numbers and Phenotypic Ratios

Given the assumptions described in the previous example, the number of each phenotype produced by the hybrid can be estimated if the number of gene pairs involved is known, and if the number of progeny is large enough to expect all genotypes to occur due to chance. Those assumptions, again, are that each additive gene contributes the same; effects of dominance, epistasis, and linkage do not exist; and environmental effects are absent or under total control. If the phenotypic expectations can be predicted by knowing the number of pairs of genes involved, it follows that if the ratio and number of phenotypes is known, the number of genes involved can be estimated.

With polygenic inheritance, the phenotypic ratio of a cross among hybrids for n number of genes can be represented by the coefficients of the binomial expanded to the nth power. In the preceding example, two pairs, or four genes, were involved. Expanding $(p + q)^4$ produces the coefficients 1, 4, 6, 4, and 1. If six genes (three pairs) were involved, the ratio of phenotypes expected from the trihybrid cross would be 1 : 6 : 15 : 20 : 15 : 6 : 1. The sum of the numbers in the ratio in this case is 64. This means that with a polygenic trait controlled by three pairs of genes, to expect one individual of each extreme the number of F_2 progeny must be approximately 64. The greater the number of genes involved, the larger the population needed to expect one of either extreme phenotype to occur (see Table 10.2).

With regard to dihybrid ratios, another has been added to the many kinds that have been explained in previous chapters. Just as dihybrid cross ratios of 9 : 3 : 3 : 1, 3 : 6 : 3 : 1 : 2 : 1, and 9 : 3 : 4 revealed to us how genes were interacting, so does the existence of the 1 : 4 : 6 : 4 : 1 ratio.

10.3 PHENOTYPIC EFFECTS OF GREATER NUMBERS OF GENES

10.3.1 An Example from Chickens

Let us assume that mature size in chickens is controlled by four pairs of genes with additive and cumulative effects. We shall make the same assumptions that were made with respect to the example of spotting in cattle.

TABLE 10.2   Theoretical Expectations in F_2 Populations with Respect to Traits Controlled by Polygenic Inheritance

Number of Pairs of Alleles	Size of F_2 Necessary for All Combinations	Number of Phenotypic Classes	Chance of Obtaining Either Extreme
1	4	3	$\frac{1}{4}$
2	16	5	$\frac{1}{16}$
3	64	7	$\frac{1}{64}$
4	256	9	$\frac{1}{256}$
5	1024	11	$\frac{1}{1024}$
6	4096	13	$\frac{1}{4096}$
n	4^n	$2n + 1$	$\frac{1}{4^n}$

In reality, this trait is probably controlled by several more than four pairs of genes, and some of the earlier assumptions may not be correct. However, for purposes of this illustration, the example will be quite satisfactory.

The additive or contributing genes will be represented by the letters *A, B, C,* and *D.* The neutral or noncontributing alleles will be represented by *a, b, c,* and *d,* respectively. Each additive gene will be assumed to contribute 75 grams (g) to the mature weight. The genotype *aabbccdd,* no additive genes, will be assumed to be that of a 750-g bird. Since the average weights of males and females are not expected to be the same, a further assumption is made that all weights will be adjusted to a male basis. The number of phenotypes will be determined by the number of additive genes present in the genotypes, zero through 8, a total of 9. The numbers representing the theoretical phenotypic ratio of the tetrahybrid cross (*AaBbCcDd* × *AaBbCcDd*) will be equal to the coefficients of the binomial expanded to the eighth power. The results are summarized in Table 10.3.

Note that in this example, very little attention has been paid to genotypes. Because of the additive relationship among the genes and the absence of dominance between alleles, a number of different genotypes can produce the same or similar phenotype. For example, a 1050-g bird could be represented by any one of 70 genotypes, and for the most part, we would not be concerned as to which one it might be. The phenotypes become much more important than any specific genotype.

TABLE 10.3   Polygenic Inheritance of Mature Size in Chickens

Number of Additive Genes	Phenotype	Number of Phenotypes
0	750 g	1
1	825 g	8
2	900 g	28
3	975 g	56
4	1050 g	70
5	1125 g	56
6	1200 g	28
7	1275 g	8
8	1350 g	1
		256

10.3.2 Normal Distribution

If the coefficients of the expanded binomial are plotted on graph paper and the points connected by a smooth line, the resulting curve conforms very well to the shape of the curve of normal distribution, the well-known "bell-shaped" population curve. Figure 10.1 is a histogram of the data contained in Table 10.3, with a normal curve superimposed.

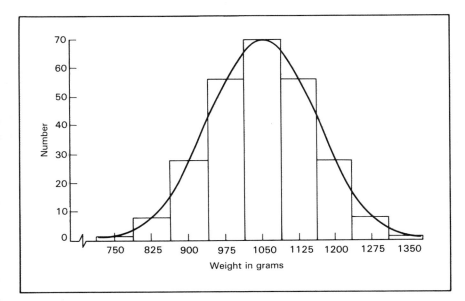

Figure 10.1   Histogram of chicken weights from data in Table 10.3, with curve of normal distribution superimposed.

If a histogram were prepared of a trait affected by a greater number of genes, there would be more bars in the histogram, and the width of each bar would be narrower. The number of points on the curve would be greater and closer together. Figure 10.2 contains four histograms to help illustrate this point. Letting each bar on a histogram represent a separate phenotype, as the number of phenotypes increases, the increments between the phenotypes become smaller and smaller. As the number of phenotypes increase and the increments between them decrease, at some point the difference between any two phenotypes will become indistinguishable without the aid of a set of scales or other appropriate measuring device. These examples should enhance our understanding and appreciation of the concept of quantitative inheritance and continuous variation.

10.3.3 Transgressive Variation

Suppose that some hens from a breed of chicken with a mean weight for cocks of 750 g were bred to some cocks from a large breed with a mean weight of 1350 g. Since these are pure breeds, it would be assumed that they were homozygous (*aabbccdd* × *AABBCCDD*). The F_1 progeny would be tetrahybrids and be expected to weigh an average of about 1050 g. Crossing the hybrids among themselves would produce F_2 progeny in a proportion similar to that listed in Table 10.3. However, when the experiment was actually conducted, assume that some birds were produced that matured at

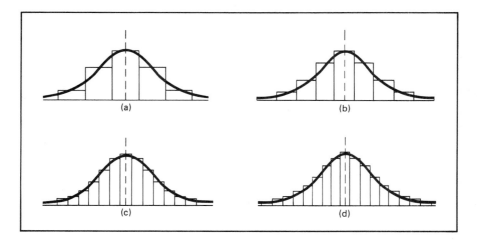

Figure 10.2 Histograms and curves of normal distribution for four populations: (a) two pairs of genes; (b) four pairs of genes; (c) six pairs of genes; and (d) eight pairs of genes.

weights larger than the largest of the parental breeds, while others matured at weights that were smaller than the smallest breed. Suppose, for example, that the extremes were 600 g and 1500 g. This phenomenon, where the extremes observed among the F_2 progeny exceed the extremes observed among the P_1, is called *transgressive variation.*

Since the members of the small breed of chicken were not as small as some of the smallest of the F_2 generation, they probably carried a pair of additive genes. Similarly, since the members of the largest breed were not as large as some of the largest of the F_2 generation, they probably carried a pair of neutral genes. One explanation is that the trait might be controlled by five pairs of genes rather than four. The probable genotype of the small breed might be *aabbccddEE,* while that of the large breed might be *AABBCCDDee.* The hybrids produced from crossing these two breeds would be expected to be intermediate in weight as they were, but heterozygous for five pairs of genes rather than four (*AaBbCcDdEe*). Assuming that the F_2 population were large enough, somewhere in the neighborhood of 1000 birds (see Table 10.2), at least one bird would be expected of the genotype *aabbccddee,* and one of the genotype *AABBCCDDEE.* Following the assumption that each additive gene is worth about 90 g, the smallest bird would weigh 600 g; the largest would weigh 1500 g.

10.4 OTHER EXAMPLES OF POLYGENES

In addition to the few relatively simple examples used so far in this chapter to illustrate polygenic inheritance, many more examples can be given. Most of those traits are probably controlled by many more pairs of genes than are illustrated in our examples. Most of the traits probably do not fully conform to the list of assumptions that were made in the theoretical explanation of the principles of polygenic inheritance. A trait such as milk production in dairy cattle, for example, may be controlled by hundreds of pairs of genes. The actual number of genes is unknown, and based on comments made earlier, it probably is not critical to know exactly how many genes are involved.

Other examples of traits controlled by multiple genes are growth rates prior to and after weaning, fertility traits such as calving and foaling percentages, litter size and hatchability, carcass traits in meat animals, speed in race horses, and hunting instinct in dogs. With so many genes affecting each trait, some examples of degrees of dominance and epistasis may exist; it is very likely that several linkage groups may be involved; multiple alleles may exist at some loci; and it is essentially a certainty that most quantitative traits are affected in varying degrees by environment. Generally, the variation observed within populations for quantitative traits exhibits characteristics similar to those that have been described in the preceding examples.

Much has been presented in earlier chapters concerning the inheritance of colors in a number of domestic animals. Another genetic factor affecting color variation is undoubtedly that of polygenes. Various shades and degrees of intensity of color, as well as the wide variation in the expression of amount of white in spotted animals has been attributed in part to polygenes by a number of genetic authorities. Examples include the variation in intensity of red in Hereford cattle, the range of shades in chestnut in horses, and the degrees of expression of yellow and orange in dogs and cats.

STUDY QUESTIONS AND EXERCISES

10.1. Define and distinguish between qualitative and quantitative inheritance.

10.2. Compare the definitions of *continuous variation* and *discontinuous variation*.

10.3. Compare the meanings of the terms *additive gene action* and *nonadditive gene action*.

10.4. Generally, what type of allelic interaction is observed among the various genes involved in polygenic inheritance?

10.5. Explain why we might be less concerned about actual genotypes when studying polygenic inheritance compared to traits affected by dominance and epistasis.

10.6. Explain the concept of transgressive variation.

PROBLEMS

10.1. Among 1000 individuals chosen at random from a certain population of mice, the following numbers and weights of each were observed:

1	23–27 g
11	28–32 g
43	33–37 g
114	38–42 g
207	43–47 g
244	48–52 g
205	53–57 g
118	58–62 g
47	63–67 g
9	68–72 g
1	73–77 g
1000	total

(a) Using letters of your choice, write the probable genotype of the mouse in the 23- to 27-g group. (b) Write the genotype of the mouse in the 73- to 77-g group. (c) Assume that a male and female mouse each weighing 35 g are chosen from a certain inbred strain and allowed to mate. Among several F_2 litters, all mice reaching maturity weighed in the range 33 to 37 g, what are the probable genotypes of the two 35-g parents? (d) A male mouse from an unrelated strain of 35-g mice was mated to several females of the original 35-g strain. Among these F_2 progeny, there were some mice as small as 23 and 24 g and some as large as 46 and 47 g; how can this be explained?

10.2. Assume that backfat thickness in swine is controlled by three pairs of additive genes. Suppose that each additive gene contributes 4 millimeters (mm) to backfat thickness and that the overall population mean is 28 mm. (a) How thick would you expect the fat to be on the hogs with the thinnest and thickest layers of fat, respectively? (b) Indicate the genotypes of the hogs with the thinnest and thickest layers of fat. (c) How many phenotypic classes should exist? (d) How large a population would be required to expect at least one pig with the thinnest possible fat layer, and one with the thickest? (e) Assume that several hogs with 28 mm of backfat that are thought to be heterozygous at all three loci are mated among themselves; if all possible combinations of genotypes are produced, how many pigs of each phenotype will be expected among the progeny? (f) Assume that when the cross in part (e) was carried out, about 250 pigs were produced in all litters. When they matured and backfat probes were taken, the population mean had not changed, and the distribution was similar to what was predicted. However, there were a very few animals that had less than the amount of backfat expected and a few with more than what was expected; what term describes this occurrence? (g) In light of the results described in part (f), how many genes would you say are affecting backfat thickness?

REFERENCES

BURNS, G. W. 1980. *The Science of Genetics,* 4th ed. Macmillan Publishing Co., Inc., New York. Chapter 9.

HUTT, F. B., and B. A. RASMUSSEN. 1982. *Animal Genetics,* 2nd ed. John Wiley & Sons, Inc., New York. Chapter 12.

SNYDER, L. H., and P. R. DAVID. 1957. *The Principles of Heredity.* D. C. Heath and Company, Lexington, Mass. Chapter 14.

ELEVEN

Variation
and Statistics

Variation refers to the differences that are observed among individuals within a population. The mean birthweight of calves in one herd of cattle may be 75 lb, but the weights of individual calves could vary from 50 lb to over 100 lb. In another herd with the very same mean (75 lb), the individual weights may range from a low weight of 65 lb to a high weight of 85 lb. The variations within the two populations are clearly different. Statistics is a branch of mathematics that deals with the collection, analysis, and interpretation of large amounts of data.

11.1 POPULATIONS AND SAMPLES

A *population* includes all animals within a large group or geographical area. It may refer to all members of a species or breed within a defined area. The limits of a population must usually be predefined. *Population genetics* refers to the genetic makeup of a population. Since a population is normally very large, it may not be possible to examine every member of the population to determine their genetic characteristics. The characteristics of a small group of animals may be used as a measure or indicator of the characteristics of the larger population. This smaller group is called a *sample*.

One important characteristic of a population or sample is the mean. The population mean is calculated by summing the values of all individuals in a population and dividing by the number of individuals in the population. It is often called the *arithmetic mean* or simply the *average* of the population. The Greek lowercase letter mu (μ) is usually used to represent the population mean. The mathematical formula is $\mu = \Sigma\ X/n$, where Σ means to add, X represents the values of each observation in the population, and n is the number of individuals.

In most cases, the population will be so large that it is impractical, or even impossible, to obtain a measurement of every individual in the population. In these instances, the mean of a sample is used to estimate the population mean. The symbol for the sample mean is $\bar{x}$. The mathematical formula for the sample mean is the same as for the population mean except that it applies to a smaller group. The individuals used to determine the sample mean should be chosen at random so that the sample will better represent the population. The size of the sample is important in that the larger the sample size, the better it will represent the population.

11.2 MEASUREMENT OF VARIATION

Determining the differences between two populations is more than observing that their two means may appear to be different. It is important to have a reliable measure of the amount of variation within the two populations. Two common measures of variation used in statistics are the variance and standard deviation.

11.2.1 Statistical Variance

Variance is a measure of the amount of variation in a population. The larger the variance, the larger the variation. The symbol for the statistical population variance is σ^2 (Greek lowercase letter sigma). The formula for calculating the population variance is

$$\sigma^2 = \frac{\Sigma\ (X - \mu)^2}{n} \tag{1}$$

where X represents values of each individual in the population, μ is the population mean, and n is the number of individuals in the population. As in the case with the population mean, the population variance is seldom determinable, so the variance of a random sample is obtained. The formula for sample variance is

$$s^2 = \frac{\Sigma\ (X - \bar{x})^2}{n - 1} \tag{2}$$

The major difference between this formula and the one for population variance [formula (1)] is that rather than dividing the sum of squared deviation by n, the number $n - 1$ is used. In other words, a degree of freedom is lost. Table 11.1 illustrates how the mean and variance are calculated.

TABLE 11.1   Calculation of Mean and Variance of Pig Weights (Pounds)

(1) Pig Number	(2) Weaning Weights	(3) Deviations from Mean	(4) Squared Deviations	(5) Square of Weights
1	17	+2	4	289
2	14	−1	1	196
3	14	−1	1	196
4	19	+4	16	361
5	16	+1	1	256
6	11	−4	16	121
7	15	0	0	225
8	16	+1	1	256
9	13	−2	4	169
10	15	+0	0	225

$$\Sigma\,X = 150 \qquad \Sigma\,(X - \bar{x})^2 = \qquad 44 \qquad \Sigma X^2 = 2294$$
$$\bar{x} = \quad 15 \qquad\qquad\qquad\qquad s^2 = \quad 4.9 \quad (\Sigma\,X)^2/10 = 2250$$
$$s = +2.2 \qquad \Sigma\,x^2 = \quad 44$$

The concept of *degrees of freedom* may be a little difficult to understand for people who are not statisticians, which includes most of us. Although it will not fully explain the use of the principle of degrees of freedom, an example may help. Suppose that you have three coins in your hand, say a nickel, a dime, and a quarter. If you allow someone to choose any coin, they are free to choose any one of the three. After the first is chosen, and someone is given an opportunity to choose a second coin, they still have freedom of choice between the two remaining coins. With only one coin remaining, anyone given the opportunity of choosing a coin has no options; they must choose the one that is there. So, among the three opportunities to choose a coin, only two of them provided any freedom of choice. The number of degrees of freedom was one less than the number of opportunities to choose, that is, $n - 1$.

Formula (2) is useful with relatively small samples and values of small magnitude. With larger samples and larger numbers, and when using calculators and computers, formula (3) is more useful. It does not require that the mean or individual deviations be calculated.

$$s^2 = \frac{\Sigma X^2 - \dfrac{(\Sigma X)^2}{n}}{n - 1} \tag{3}$$

This procedure is also illustrated in Table 11.1. Note that the sum of the squared deviations at the bottom of column 4 is the same value signified by the symbol $\Sigma\ x^2$ at the bottom of column 5. This value is often called the *mean of the squares,* or simply the *mean square.* It is the value that when divided by $n\ -\ 1$ produces the variance. Formula (3) is often called the *machine formula* or the *computational formula* for calculating the variance.

Note that units of variance are expressed in squared terms. In the example in Table 11.1, the values of weights are expressed in pounds; the units of variance are lb^2. To most readers, this is understandably difficult to comprehend. It might be easier to accept if we think of the variance as an index of variation rather than something that can be weighed or measured. As we will see in later examples, the variance is a very valuable statistic.

11.2.2 Standard Deviation

The *standard deviation* is a measure of the average deviation of observations from the sample or population mean. The units are the same as those of the mean. Simply stated, the standard deviation is the square root of the variance. The symbol for the population standard deviation is a lowercase sigma (σ). The symbol for the sample standard deviation is a lowercase *s*. The formula for the standard deviation is the same as for the variance with one additional step—the calculation of the square root. The value of *s* is shown in Table 11.1.

11.2.3 Coefficient of Variation

When comparing two populations, it is sometimes desirable to know which has the greater or lesser degree of variation. If the means of the two populations were close to the same value, it would be relatively easy to see that the one with the larger standard deviation would have the greater variation. For example, a population with a mean of 50 and a standard deviation of ± 15 would have less variation than a population with a mean of 50 and a standard deviation of ± 24.

Suppose that two populations have means and standard deviations of $27\ \pm\ 9$ and $43\ \pm\ 11$, respectively. To compare the relative variation of these two populations, calculate the coefficient of variation for each. The coefficient of variation is simply the ratio of the standard deviation to the mean, or the amount of variation per unit of mean, or, more simply, the result of dividing the standard deviation by the mean. The formula for the coefficient of variation is

$$\text{C.V.} = \frac{s}{\bar{x}} \tag{4}$$

The coefficient of variation of the first population is 9 ÷ 27, or 0.333. The second population has a coefficient of variation of 11 ÷ 43, or 0.256. It is now easy to see that the first of the two populations has the greater variation.

11.3 CHARACTERISTICS OF A NORMAL DISTRIBUTION

The data in Table 11.2 represent the heights of a sample of 532 adult male college students. This is a good example of a quantitative trait as described in Chapter Ten. When samples are sufficiently large, observed values of most quantitative traits tend to fit a normal distribution. Figure 11.1 shows a histogram of the data in Table 11.2, with a standard normal curve superimposed. A perpendicular line extended upward from the base at the point of the mean will intersect the curve at its highest point. The height of the curve at any point represents the number of individuals with that particular value. Values nearest the mean are represented by the greatest frequencies. As values deviate more and more from the mean, their frequencies become less and less. The total area under the curve represents the total number of individuals in the population.

TABLE 11.2   Height in Inches of a Sample Population of 532 Adult College Men

Height, X	Frequency f	Height times Frequency	Deviation, d	d^2	$d^2 \times f$
64	2	128	7	49	98
65	4	260	6	36	144
66	13	858	5	25	325
67	23	1541	4	16	368
68	51	3468	3	9	459
69	57	3933	2	4	228
70	73	5110	1	1	73
71	88	6248	0	0	0
72	81	5832	1	1	81
73	51	3723	2	4	104
74	47	3478	3	9	423
75	21	1575	4	16	336
76	14	1064	5	25	350
77	3	231	6	36	108
78	4	312	7	49	196
	$n = 532$	$\Sigma\ X = 37{,}761$			$\Sigma x^2 = 3293$
		$\bar{x} = 71.0$			$s^2 = 6.2$
					$s = 2.5$

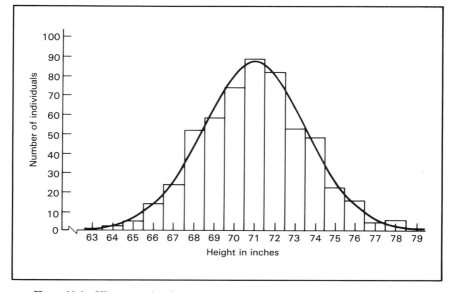

Figure 11.1 Histogram showing variation in heights among the population of adult men listed in Table 11.2. A standard normal curve is superimposed.

The nature of the normal distribution is such that the areas under various segments of the curve can be calculated and predicted. If a perpendicular line is extended from a point on the base one standard deviation to the left of the mean upward to intersect the curve, and another such line is drawn one standard deviation to the right of the mean, the area under the curve between these two lines would represent 68.26 percent, or about two-thirds of the total population. For the data in Table 11.2, this means that approximately 360 men in the sample population should have heights between 68.5 and 73.5 inches. The data in the sample between these two values actually number 350. Note from the histogram in Figure 11.1 that although the data do not fit a normal distribution perfectly, they are reasonably close.

Two standard deviations on either side of the mean of a standard normal population will include 95.44 percent (or roughly 95 percent) of that population. Three standard deviations on either side of the mean will include 99.74 percent of a normal population. For all practical purposes, especially in populations of several hundred or fewer individuals, this is essentially the entire population.

Table 11.3 shows percentages of a normal population included within given ranges of standard deviations from the mean. For any given percent-

**TABLE 11.3   Areas under a Normal Curve for Given Units
of Standard Deviations on Either Side of the Mean**

Number of Standard Deviation Units on Each Side of the Mean	Percent of the Curve Included
0.5	38.3
0.675	50.0
0.84	60.0
0.94	65.0
1.0	68.3
1.04	70.0
1.15	75.0
1.28	80.0
1.5	86.6
1.645	90.0
1.96	95.0
2.0	95.4
2.33	98.0
2.5	98.8
2.585	99.0
3.0	99.74
3.5	99.96
3.9	99.999
4.0	100.00

age of a population within certain specified limits, 100 minus the percentage would be the portion lying outside, or beyond, those limits under the two "tails" of the curve. For example, approximately 95 percent of a normal population lies between two standard deviations on either side of the mean; therefore, approximately 5 percent lies outside these limits. Further, since the curve is symmetrical, $\frac{1}{2}$ of the 5 percent (or 2.5 percent) lies under the left tail and another 2.5 percent lies under the right tail. Figure 11.2 shows diagrammatically the areas of several segments of a normal curve as well as the areas of several combined segments.

11.4 METHODS OF STATISTICAL ANALYSIS

Many applications of statistical procedures can be made in genetics, but only a few will be presented here, and they will be simplified as much as possible. The procedures that will be explained are regression and correlation analyses, and statistical comparison of two or more sample means. Some of the uses and importance are presented in this and later chapters.

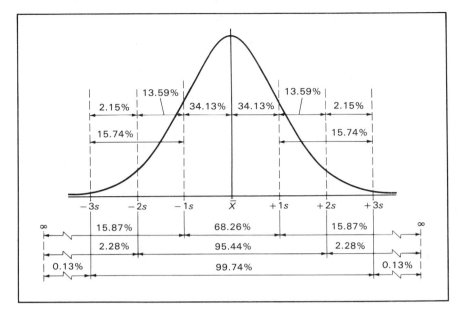

Figure 11.2   Predicted percentages of a normal population expected to occur within specified segments of that population.

11.4.1 Comparing Two Sample Means

When comparing the means of two sample populations, it may be desirable to know whether the means are truly different, or whether they are different due to sampling error. One statistic used to help make this decision is called the *standard error of the difference between means* (s_d). The formula is

$$s_d = \sqrt{\frac{s_1^2}{n_1} + \frac{s_2^2}{n_2}} \tag{5}$$

where s_1^2 and s_2^2 are the variances of the two samples being compared.

To illustrate, suppose that observations were made of weaning weights of a second sample of 10 pigs similar to the one in Table 11.1, which had a mean and variance of 15 and 4.9, respectively. Assume that the mean and variance of the second sample were 16.5 and 5.8, respectively. Using formula (5), the standard error of the difference between the two means would be

$$S_d = \sqrt{\frac{4.9}{10} + \frac{5.8}{10}} = \pm 1.03$$

To determine the significance of the difference between the two sample means, that difference is compared to s_d. The ratio of the difference between means compared to s_d is calculated using the expression

$$\frac{x_1 - x_2}{s_d} \qquad (6)$$

The difference between means and s_d is expressed as positive numbers. In the example above, the ratio of the difference between means and s_d is $(16.5 - 15)/1.03$, or 1.46. Recall that the area under a normal population curve representing one standard deviation on either side of the mean includes about 68 percent of the population. The area represented by two standard deviations about the mean includes about 95 percent of the population. In this example, the difference between means is equal to 1.46 standard deviation units (standard error units in this case), or about 1.5 units. Referring to Table 11.3, it can be seen that the area under a normal curve, including 1.5 standard deviations about the mean, represents 86.6 (or about 87) percent of the population. This would indicate that about 13 $(100 - 87)$ percent of the population would be expected to exceed the limits of 1.5 standard deviations about the mean.

If a conclusion is made that the difference between the two means, 16.5 and 15, is due to some factor other than chance (in other words, two populations are involved), we can be about 87 percent confident that it is the correct conclusion. Stating it another way, if the conclusion is drawn that two populations are involved, the probability that it is the wrong conclusion is about 13 percent.

Many scientists hold the position that to assure that the difference between two means is statistically significant, the probability of an error should be around 5 percent or less. In other words, they would like to make a decision with at least a 95 percent level of confidence that it was correct. To require this level of confidence in a decision that two sample means are truly different, the ratio of the difference between means and s_d would need to be about 2. In this example, it was about 1.5; therefore, most statisticians would conclude that the two means are probably from the same population and that their difference was due to sampling error. It should be stated that the use of statistics does not prove any decision to be correct or incorrect. It merely shows the probabilities of being right or wrong if a certain conclusion is drawn. It is another of the many tools that scientists can use in the decision-making process.

11.4.2 Analysis of Variance

There are occasions when an experiment may involve three or more samples. In these instances, the procedure presented previously for comparing two sample means will not apply. The analysis of variance (ANOVA) can be used for this purpose. In some situations it may also be used for comparing two sample means.

As the name suggests, this procedure actually compares variances. Variation between or among samples is compared to the variation within the samples. The ratio of these two sources of variation is referred to as the *F ratio*. Tables have been prepared that allow scientists to determine the levels of significance of such ratios. Generally, the variation among samples is a function of the differences among the means, so the greater the differences among sample means, the greater the variation among samples, and the greater the *F* ratio. So even though the procedure calls for the comparison of variances, it is used to determine the significance of differences between two means and among three or more means.

Probably the best way to illustrate how the analysis of variance works is to use a simple example. Let us use a hypothetical situation involving three samples of 6 from a large population. The three samples might represent the effect of three treatments in an experiment to determine the proper requirements for a particular vitamin on weight gains in baby chicks, or any of a number of other hypothetical examples. The data from the experiment are shown in Table 11.4.

The first step in the analysis is to obtain a measure of variation for the total population, including all observations in each sample. The procedure makes use of the numerator of the so-called computational or machine formula for calculating the variance [formula (3)]: $\Sigma\,(X^2) - \Sigma\,X^2/n$, where the X's represent individual observations, and n is the total number of observations in the experiment. The figure produced by applying the formula to the data in the table represents the sum of all squared deviations from the mean. The same result would have been obtained if the formula

TABLE 11.4   Comparison of Three Hypothetical Sample Means by Analysis of Variance

Treatment A	Treatment B	Treatment C
6	5	7
5	7	6
6	6	6
4	7	8
7	5	5
5	8	8
$\Sigma_A = 33$	$\Sigma_B = 38$	$\Sigma_C = 40$
	Total ($\Sigma\ X$'s) = 111	
	$n = 18$	

$\Sigma (X - \bar{x})^2$ [the numerator of formula (2)] had been applied. The result is usually referred to as *total sum of squares,* or total SS, or SS_t:

$$\frac{(6^2 + 5^2 + 6^2 + \cdots + 8^2 + 5^2 + 8^2) - 111^2}{18} = 709.0 - 684.5 = 24.5$$

The next step is to calculate the variation among treatments, called *treatment sums of squares,* or *sums of squares among treatments.* This value is determined by squaring each treatment sum, dividing by the number of observations in the treatment, then summing those values, and finally, subtracting the so-called correction factor $\Sigma X^2/n$:

$$\text{treatment SS} = \frac{\Sigma_A^2}{n_A} + \frac{\Sigma_B^2}{n_B} + \frac{\Sigma_C^2}{n_C} - \frac{\Sigma X^2}{n}$$

$$= \frac{33^2}{6} + \frac{38^2}{6} + \frac{40^2}{6} - \frac{111^2}{18}$$

$$= (181.5 + 240.7 + 266.7) - 684.5 = 4.4$$

Among-treatment sums of squares can be referred to as SS_{among} or simply SS_a.

The third step is to determine the *within-treatment sums of squares* or SS_w. This procedure involves merely calculating the difference between the total SS and treatment SS:

$$SS_w = SS_t - SS_a$$
$$= 24.5 - 4.4 = 20.1$$

Dividing the among- and within-treatment sums of squares by their respective degrees of freedom produces the among- and within-treatment variances, or mean of squares or simply mean squares, MS_a and MS_w, respectively. The simplest way to illustrate these calculations is to set up a table called the *analysis of variance* (ANOVA) *table* (see Table 11.5).

Total degrees of freedom for the experiment are $n - 1$ or 17. The among-treatment degrees of freedom are equal to the number of treatments minus 1, in this case $3 - 1 = 2$. The degrees of freedom within treatments are the difference between the total number of observations in the experi-

TABLE 11.5   Analysis of Variance of Data in Table 11.4

Source of Variation	SS	Degrees of Freedom	MS	F Ratio
Among treatments	4.4	2	2.20	1.65
Within treatments	20.1	15	1.33	
Total	24.5	17		

ment and the number of treatments, in this case $n - 3 = 15$. Notice in the table that the number of degrees of freedom within is the difference between the total degrees of freedom and degrees of freedom among treatments, $17 - 2 = 15$.

The final step in the calculations is to determine the F ratio, which is the ratio of the among- and within-treatment variances. If these variances (mean squares) are the same or very similar in magnitude, the F ratio will approach the value of 1. In other words, an F ratio of close to 1 suggests little or no difference between variances, and in turn little or no differences among the various treatment means. In this example experiment, the F ratio is 1.65, indicating that the among-treatment MS is a little larger than the within-treatment MS. Most scientists would probably conclude that these variances are not different enough to be statistically significant.

Tables have been constructed to help determine levels of significance for F ratios (see Table 11.6). The larger the F ratio, the greater the probability that the variances are different, and the more significant are the differences among sample or treatment means. In this example, for differences to have been significant at the 5 percent level of probability, the F ratio would have to be at least as large as 3.68. Significance at the 1 percent level would require an F ratio to be at least 6.36.

To find the appropriate F ratio in the table, use the degrees of freedom associated with the two mean squares. The among-treatment mean square is associated with two degrees of freedom. Find the column headed by the number 2 under either the 0.05 or 0.01 level of probability sections of the table. Move down that column to the number in the row pertaining to the degrees of freedom associated with the within-treatment mean square.

11.4.3 Regression Analysis

A *regression coefficient* represents how much one variable, X, can be expected to change per unit of change in another variable, Y. A series of X and Y coordinates plotted on graph paper is called a *scatter diagram*. A straight line drawn among the plotted points that represents the best relationship between pairs of X and Y values is called a *regression line*. Data on birth weights of cows (X) and their calves (Y) in columns 2 and 5 of Table 11.7 are plotted in Figure 11.3.

The formula for a straight line, called a *linear regression line,* takes the form $Y = a + bX$, where Y is some predicted value, a is the Y-intercept (the point where the regression line crosses the Y axis), b is the slope of the line, and X is the value of a second variable. For given values of a and b, the Y value depends on the value of X; therefore, Y is called the dependent variable and X the independent variable.

TABLE 11.6 F Ratios at the 5 and 1 Percent Levels of Significance

| | Degrees of Freedom among Groups or Treatments | | | | | | | | | | | | |
| Degrees of Freedom Within | 0.05 Level | | | | | | 0.01 Level | | | | | |
	1	2	3	4	5	6	1	2	3	4	5	6
5	6.61	5.79	5.41	5.19	5.05	4.95	16.3	13.3	12.1	11.4	11.0	10.7
6	5.99	5.14	4.76	4.53	4.39	4.28	13.7	10.9	9.78	9.15	8.75	8.47
7	5.59	4.74	4.35	4.12	3.97	3.87	12.2	9.55	8.45	7.85	7.46	7.19
8	5.32	4.46	4.07	3.84	3.69	3.58	11.3	8.65	7.59	7.01	6.63	6.37
9	5.12	4.26	3.86	3.63	3.48	3.37	10.6	8.02	6.99	6.42	6.06	5.80
10	4.96	4.10	3.71	3.48	3.33	3.37	10.0	7.56	6.55	5.99	5.64	5.39
11	4.84	3.98	3.59	3.36	3.20	3.09	9.65	7.21	6.22	5.67	5.32	5.07
12	4.75	3.89	3.49	3.26	3.11	3.00	9.33	6.93	5.95	5.41	5.06	4.82
13	4.67	3.81	3.41	3.18	3.03	2.92	9.07	6.70	5.74	5.21	4.86	4.62
14	4.60	3.74	3.34	3.11	2.96	2.85	8.86	6.51	5.56	5.04	4.70	4.46
15	4.54	3.68	3.29	3.06	2.90	2.79	8.68	6.36	5.42	4.89	4.56	4.32
16	4.49	3.63	3.24	3.01	2.85	2.74	8.53	6.23	5.29	4.77	4.44	4.20
18	4.41	3.55	3.16	2.93	2.77	2.66	8.29	6.01	5.09	4.58	4.25	4.01
20	4.35	3.49	3.10	2.87	2.71	2.60	8.10	5.85	4.94	4.43	4.10	3.87
25	4.24	3.39	2.99	2.76	2.60	2.49	7.77	5.57	4.68	4.18	3.86	3.63
30	4.17	3.32	2.92	2.69	2.53	2.42	7.56	5.39	4.51	4.02	3.70	3.47
40	4.08	3.23	2.84	2.61	2.45	2.34	7.31	5.18	4.31	3.83	3.51	3.29
60	4.00	3.15	2.76	2.53	2.37	2.25	7.08	4.98	4.13	3.65	3.34	3.12
120	3.92	3.07	2.68	2.45	2.29	2.18	6.85	4.79	3.95	3.48	3.17	2.96
∞	3.84	3.00	2.60	2.37	2.21	2.10	6.63	4.61	3.78	3.32	3.02	2.80

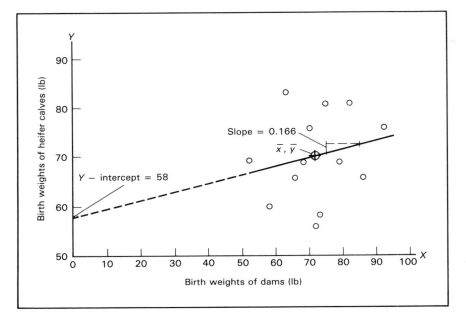

Figure 11.3 Regression of birth weights of heifer calves on the birth weights of their dams.

The slope (b) of the regression line is also called the *coefficient of regression* of Y on X. The formula for calculating b is

$$b = \frac{\Sigma XY - \dfrac{(\Sigma X) \cdot (\Sigma Y)}{n}}{\Sigma X^2 - \dfrac{(\Sigma X)^2}{n}} \tag{7}$$

The denominator in this formula is the same as the numerator for calculating variance. The numerator of formula (7) is referred to as the sum of cross products, S_{xy}, at the bottom of column 6 of Table 11.7. The value of b is the ratio of the sum of cross products (S_{xy}) to the sum of squares for variable X (SS_x). In this example, the regression coefficient (b) is 247/1488 = 0.166. Again, this coefficient indicates that for every 1-lb change in birth weights of cows, the change in the birth weights of daughters was 0.166 lb, or about $\frac{1}{6}$ lb.

In order to draw the regression line on graph paper, two points are needed; one is at the coordinates of the means of the two variables X and Y; the other is the Y-intercept. For the example in Table 11.7, the two means

TABLE 11.7   **Data for Performing a Regression Analysis of Pounds of Weight at Birth of Heifer Calves on Birth Weights of Their Dams**

(1) Cow Number	(2) Birth Weights of Cows, X	(3) Squares of Cow Weights, X^2	(4) Birth Weights of Calves, Y	(5) Squares of Calf Weights, Y^2	(6) Cross Products of Birth Weights, XY
1	66	4356	66	4356	4356
2	52	2704	69	4761	3588
3	86	7396	66	4356	5676
4	75	5625	81	6561	6075
5	70	4900	76	5776	5320
6	68	4624	69	4761	4692
7	82	6724	81	6561	6642
8	58	3364	60	3600	3480
9	72	5184	56	3136	4032
10	79	6241	69	4761	5451
11	73	5329	58	3364	4234
12	92	8464	76	5776	6992
13	63	3969	83	6889	5229
	936	68,880	910	64,658	65,767

$$\frac{936^2}{13} = 67,392 \qquad \frac{910^2}{13} = 63,700 \qquad \frac{936 \times 910}{13} = 65,520$$

$\bar{x} = 72$ $SS_x = 1488$ $\bar{y} = 70$ $SS_y = 958$ $S_{xy} = 247$

$Var_x = 124$ $Var_y = 79.8$ $Cov_{xy} = 20.6$

$s_x = 11.1$ $s_y = 8.9$

are presented at the bottoms of columns 2 and 4. Plotting this coordinate on a graph provides one point through which the regression line will pass.

The value of a, the Y-intercept, can be calculated using the formula $a = \bar{y} - b\bar{x}$, where $\bar{y}$ and $\bar{x}$ are the means of all Y values and X values, respectively, 70 and 72 in this example. The value of the Y-intercept is

$$a = 70 - (0.166 \times 72) = 70 - 12 = 58$$

Another method of calculating the slope of the regression line (b) is to make use of the covariance and the variance of X. If the numerator of the regression formula (7) is divided by $n - 1$ (n being the number of observations per cow in this example), the result is called the covariance (Cov_{xy} in Table 11.7). If the denominator of the regression formula is divided by $n - 1$ (12 in this example), the result is the variance of X (Var_x in Table 11.7). An alternate formula for determining the regression coefficient can be expressed as the ratio of the covariance to the variance of X:

$$b = \frac{\text{Cov}_{xy}}{\text{Var}_x} = \frac{20.6}{124} = 0.166 \tag{8}$$

11.4.4 Correlation Analysis

The *correlation coefficient, r,* is a measure of how closely one set of data is related to another. It is a measure of the relative distance of X–Y coordinates from a regression line. If all points are situated on the regression line, the correlation coefficient will be $+1.0$ or -1.0, depending on whether the slope (b) is positive or negative. If b is positive, r will be positive; if b is negative, r will be negative. A correlation of 1.0 is called a *perfect correlation.* While the regression coefficient was defined as the amount of change in a variable Y for every unit of change in variable X, the correlation coefficient is defined as the change in Y in standard deviation units for every unit of change in standard deviation units for X.

The formula for calculating the correlation coefficient is

$$r = \frac{\Sigma XY - \dfrac{(\Sigma X) \cdot (\Sigma Y)}{n}}{\sqrt{\Sigma X^2 - \dfrac{(\Sigma X)^2}{n} \cdot \Sigma Y^2 - \dfrac{(\Sigma Y)^2}{n}}} \tag{9}$$

Substituting the appropriate values from Table 11.7, the correlation coefficient would be

$$r = \frac{247}{\sqrt{(1488)(958)}} = \frac{247}{1193.9} = 0.207$$

In the event that the covariance and the variances have already been calculated, an alternate formula could be used:

$$r = \frac{\text{covariance of } XY}{\sqrt{(\text{Var}_x)\,(\text{Var}_y)}} \tag{10}$$

in which case the calculation would be

$$\frac{20.6}{\sqrt{124(79.8)}} = \frac{20.6}{99.5} = 0.207$$

The relationship of b and r is such that if the variances of X and Y are identical, the values of b and r will be identical. If the value of b is known, the value of r can be calculated using the formula $r = b(s_x/s_y)$. In this example, $r = 0.166(11.1/8.9) = 0.207$. On the other hand, if the value of r is known, the value of b can be calculated using the formula $b = r(s_y/s_x) = 0.207(8.9/11.1) = 0.166$.

STUDY QUESTIONS AND EXERCISES

11.1. Define each term.

coefficient of variation	population	standard deviation
correlation coefficient	regression coefficient	statistics
covariance	sample	variance
degrees of freedom	scatter diagram	variation
mean		

11.2. Explain what it means for a population to be normally distributed.

11.3. Explain why greater deviations from the mean are likely to be observed in larger populations.

11.4. Explain how the procedure called analysis of variance can be used to analyze for difference among sample means.

11.5. What is the minimum number of points needed to draw a linear regression line?

11.6. Explain the meaning of correlation coefficient of 1.

11.7. Under what conditions will a regression coefficient and a correlation coefficient be the same for a given set of data?

PROBLEMS

11.1. A small population is represented by the following values: 23, 29, 15, 12, 25, 27, 15, 20, 13, 31, 23, 18, 22, and 21. Calculate the **(a)** mean, **(b)** variance, **(c)** standard deviation, and **(d)** coefficient of variation.

11.2. A small group of feeder lambs has the following weights: 83, 73, 79, 87, 75, 70, 79, 80, 85, 82, and 76. Calculate the **(a)** mean, **(b)** variance, **(c)** standard deviation, and **(d)** coefficient of variation.

11.3. A certain population has a mean and a standard deviation of 86.4 and 11.8, respectively. **(a)** What is the largest value that would be included within one standard deviation from the mean? **(b)** About how many standard deviation units would 72.2 be from the mean? **(c)** About what percent of the total population would be expected to have values of less than 72.2? **(d)** Between what values would you expect to find approximately 95 percent of the population? **(e)** What percent of the population would be expected to exceed the value of 115.9? **(f)** What percent of the population would be expected to exceed the limits of three standard deviations on either side of the mean? **(g)** In a sample population of 100, would any individuals be expected to express values below 51.0 or above 121.8? **(h)** Based on your answers to parts (f) and (g), about how large a sample would be needed to expect to see at least one individual each above and below the values in part (g)?

11.4. Another group of 11 lambs from the same flock as those in Problem 11.2 were raised by ewes that had been fed a special additive intended to promote milk production and thus heavier lambs. The mean of this experimental sample was 74.9 with a standard deviation of 4.9. **(a)** Calculate the standard error of the difference between the experimental and control sample means. **(b)** Calculate the ratio between the difference between means and the standard error of the difference between means. **(c)** What is the probability of error if the conclusion is made that the difference between the means was due to the effect of the additive rather than to chance? (See Table 11.3.)

11.5. Listed in the following table are data from a hypothetical experiment involving three treatments.

Treatment A	Treatment B	Treatment C
6	6	8
4	7	6
6	5	9
5	8	6
5	5	8
4	7	7

Set up an ANOVA table including total sums of squares and degrees of freedom, treatment sums of squares and degrees of freedom, within sums of squares and degrees of freedom, treatment and within mean squares, and F ratio and probability of a chance occurrence.

11.6. Following is a list of observations for two variables with an apparent relationship.

Variable X	Variable Y
1	6
2	9
3	7
4	7
5	8
7	6
8	4
9	5
10	3
10	6
11	5

Calculate **(a)** the sums of squares for X, **(b)** the sums of squares for Y, and **(c)** the sums of cross products. Calculate **(d)** the regression coefficient b, **(e)** the Y-intercept a, and **(f)** the correlation coefficient r.

REFERENCES

BENDER, F. E., L. W. DOUGLASS, and A. KRAMER. 1982. *Statistical Methods for Food and Agriculture.* AVI Publishing Company, Westport, Conn.

BURNS, G. W. 1980. *The Science of Genetics,* 4th ed. Macmillan Publishing Co., Inc., New York.

LASLEY, J. F. 1987. *Genetics of Livestock Improvement,* 4th ed. Prentice-Hall, Inc., Englewood Cliffs, N.J., pp. 431–434.

SPIEGEL, MURRAY R. 1975. *Probability and Statistics.* McGraw-Hill Book Company, New York.

STANSFIELD, W. D. 1983. *Theory and Problems of Genetics,* 2nd ed. McGraw-Hill Book Company, New York.

TWELVE

Selection
and Heritability

12.1 INTRODUCTION TO GENETIC SELECTION

The general definition of selection suggests that choices are made between or among two or more objects such that certain ones are chosen for some purpose and others are rejected. Genetic selection is a process whereby certain animals are allowed to reproduce, while others are not.

12.1.1 Natural Selection

The forces or factors that determine which members of one generation are to be allowed to be the parents of the next generation are of two kinds: natural and artificial. Natural selection involves forces of nature that determine which animals will reproduce and leave viable offspring to continue the process. Natural selection has been referred to as "survival of the fittest." Animals that are able to adapt to the environment and survive are those that are most likely to reproduce. To the extent that their ability to survive is genetically controlled, the probabilities will be increased that some of those genes will be transmitted to the next generation. This should in turn increase the chances that more members of the next generation will be better able to adapt and thus increase their chances of survival. By its very nature, changes that occur due to natural selection take place very slowly.

12.1.2 Artificial Selection

With artificial selection, humans decide which animals will be allowed to reproduce and which will not. Animals chosen to produce the next generation may not be chosen for their abilities to survive or adapt to the environment. Rather, they are chosen for their apparent superiority for some trait or combination of traits that humans have decided are useful. Since animals involved with artificial selection are usually domesticated, they are protected from many adversities of the environment. Even though the forces of natural selection continue to work, they are in large part overshadowed by the forces of artificial selection.

To the extent that we are able to determine which animals are superior for the traits we think are important, we can say it is the "best" animals that are chosen. So another rather simplified definition of artificial selection is that it is the mating of the "best to the best." Most of the material in this chapter and Chapter Thirteen will deal with this kind of selection.

12.1.3 Genetic Effect of Selection

The ultimate effect of selection is to increase the frequencies of genes that are associated with improving the productivity of domestic animals, and decreasing the frequencies of the less productive alleles. The relationship between selection and the changing of gene frequencies was explained in some detail in Chapter Seven.

12.2 HEREDITY AND ENVIRONMENT

The sources of differences observed among animals for various traits are generally of two types: genetic and environmental. A Thoroughbred stallion might be capable of winning the Kentucky Derby, partly because of a superior genetic makeup and partly because of superior training and husbandry. Both factors, genetic and environmental, are necessary. Even with his apparently superior genotype, Seattle Slew could not have won the Triple Crown without the excellent training he received. On the other hand, the best training in the world would not prepare a Belgian stallion to win against even a mediocre Thoroughbred. The example may be extreme, but it points out that both superior genotype and superior environment are necessary for animals to reach optimum performance.

12.2.1 Identifying Sources of Variation

Sources of variation can be illustrated using a mathematical equation. The letter V can be used to represent variation in general, and subscripts can be attached to indicate the source of variation. The symbol V_t will be used to represent the total variation observed for a given trait among a specified population or population sample. It might take the form of the statistical variance, or the selection differential. Total variation is also called *phenotypic variation.* The simplest equation would include the two general sources of variation, *genetic* and *environmental:*

$$V_t = V_g + V_e \qquad (1)$$

where V_g represents all types of genetic variation and V_e represents all kinds of environmental variation.

A third source of variation can be defined, one that comes through the interaction between genotype and environment. The symbol V_i can be used. A simplified example of *genetic–environmental interaction* can be taken from cattle. The continental European and British breeds of cattle were bred and selected to perform in cold and moderate climates. When they are taken to tropical and subtropical regions of the world, they do not perform as well as do cattle bred and selected for performance in tropical climates. This can happen on a smaller scale where cattle bred and developed on one farm or ranch may not do so well when moved to another ranch, where environmental conditions are quite different. Adding interaction as a source of variation, the equation becomes

$$V_t = V_g + V_e + V_i \qquad (2)$$

Genetic variation can be subdivided in a number of ways. Some genetic variation can be due to the effects of *additive* genes (V_a), also known as *polygenes, multiple genes,* and *quantitative inheritance.* Genetic effects that are not additive can be called *nonadditive* (V_n). Nonadditive variation can be subdivided into that due to dominance (V_d) and that due to epistasis (V_{ep}). The equation now expands to

$$V_t = V_a + V_d + V_{ep} + V_e + V_i \qquad (3)$$

For practical purposes, in the remaining part of this chapter we consider three subdivisions of variation: additive, nonadditive, and environmental:

$$V_t = V_a + V_n + V_e \qquad (4)$$

Environmental variation includes such factors as climate, weather, nutrition, disease, and general husbandry practices.

12.2.2 Nature of Heritability Estimates

In general, *heritability* refers to that portion of the total phenotypic variation controlled by genes. It is that part of the total variation that can be transmitted to the next generation by the gametes. Remember, again, we are dealing not with total or absolute values for traits, but with a measure of the differences among individuals within the trait. A heritability estimate is a mathematical expression of the genetic variation calculated from experimental data. It is the best estimate of the actual or true heritability that can be obtained for a given trait. Two kinds of heritability estimates can be described, heritability estimates in the *broad* sense, and heritability estimates in the *narrow* sense.

Estimates of heritability in the broad sense are calculated by expressing variation from all genetic sources as a fraction or percentage of the total phenotypic variation:

$$H^2_{\text{broad}} = \frac{V_g}{V_t} \tag{5}$$

H^2 is a standard symbol for heritability estimate. Some references use a lowercase h (h^2), but either is satisfactory. Heritability estimates of this kind include effects from both additive and nonadditive genes.

Estimates of heritability in the narrow sense include the effects of additive genes only. The formula is

$$H^2_{\text{narrow}} = \frac{V_a}{V_t} \tag{6}$$

Most methods of determining estimates of heritability in domestic animals measure the impact of only the additive genes. Hereafter, in this book, unless otherwise indicated, all references to heritability estimates will refer to those in the narrow sense. This is not to say that effects of nonadditive genes are unimportant; they manifest themselves and in different ways. Nonadditive genes, those related to dominance and epistasis, tend to affect variation in an "all or none" fashion, and are affected by environment to a lesser extent.

Heritability estimates are usually expressed as decimal fractions or as percentages. The range of estimates is zero to 1.00, or zero to 100 percent. Notice in Table 12.1 that only a few estimates are above 50 percent. It is common to group heritability estimates into three general categories, identified as low, medium, and high. Estimates in the range of zero to 20 percent are considered low. Those in the approximate range between 20 and 40 percent are medium. Highly heritable traits are those with estimates generally above 40 percent.

Traits with heritabilities in the low group include those related to fertility, such as lambing, calving, and foaling percentage; litter size in swine, dogs, and cats; and hatchability in chickens. Milk production and growth traits measured at weaning are two examples of traits with medium estimates of heritability. Highly heritable traits include those measured in animals when they are more mature, such as feedlot traits, carcass traits, and yearling and mature weights.

TABLE 12.1 Examples of Heritability Estimates of Economically Important Traits in Selected Domestic Animals

	Approximate Heritability	
Species and Trait	Percent	Generally
Swine		
Number of pigs farrowed	5–15	Low
Number of pigs weaned	10–15	Low
Litter weight at weaning	15–20	Low
Pig weight at 6 months of age	20–25	Medium
Growth rate from weaning to market	25–40	Medium
Efficiency of gain	25–40	Medium
Thickness of backfat	40–60	High
Area of the loin eye	40–60	High
Percent lean cuts	35–50	Medium to high
Beef cattle		
Calving interval	5–15	Low
Number of calves born per 100 cows bred	5–15	Low
Birth weight	25–40	Medium
Weaning weight	25–35	Medium
Yearling weight	50–60	High
Mature weight	50–60	High
Feedlot rate of gain	45–55	High
Efficiency of gains in the feedlot	40–50	High
Carcass traits in general	40–60	High
Loin eye area	40–60	High
Dairy cattle		
Milk yield	20–40	Medium
Percent fat	40–70	High
Pounds of fat	20–40	Medium
Total milk solids	20–35	Medium
Sheep		
Lambing percentage	5–10	Low
Birth weight	15–35	Low to medium
Weaning weight	10–35	Low to medium
Yearling weight	30–45	Medium to high
Mature weight	40–60	High
Feedlot rate of gain	30–40	Medium
Efficiency of gain	20–45	Medium to high
Loin eye area	30–50	Medium to high
Fleece weight	30–50	Medium to high
Fiber diameter	40–55	High
Staple length	40–60	High
Horses		
Foaling percentage	5–10	Low
Speed in Thoroughbreds	35	Medium
Trotting speed	35–45	Medium to high
Heart girth	20–30	Medium
Height at withers	25–80	Medium to high
Poultry		
Body weight	40	Medium to high
Egg weight	60	High
Number of eggs produced	20	Low to medium

12.3 DETERMINING HERITABILITY ESTIMATES

The calculation of heritability estimates is based on the principle that since relatives are more alike genetically than unrelated animals, they will tend to look more alike for a given trait than will nonrelatives. Various methods have been devised to determine mathematically the similarity between relatives to arrive ultimately at a measure of the true heritability of a trait.

12.3.1 Comparing Twins

Comparing the similarity between identical twins to the similarity between dizygotic twins has been used as one method to determine estimates of heritability in humans and cattle. Since identical, or monozygotic, twins are formed from the same fertilized ovum, their genotypes will be identical. Any differences in similarity between pairs of identical twins would be assumed to be due to effects of environment. Dizygotic twins are formed from two separate fertilized ova, so their genotypes would not be expected to be identical. Any differences between pairs of nonidentical twins would therefore be due both to effects of different genes and to environmental effects.

Suppose that in a controlled experiment, the average difference between weaning weights of identical twin calves were 45 lb. This would be a measure of environmental variation. Further suppose that the average difference between sets of nonidentical twins were 85 lb; this would be a measure of both genetic and environmental variation, or actually, total variation. Mathematically, the difference between the two values, 40 pounds, should be a measure of genetic variation. The heritability estimate would simply be the ratio between the genetic variation of 40 lb and the total variation of 85 pounds, or about 47 percent.

Estimates of heritability obtained from twin data are usually considerably higher than those obtained by other methods. It is assumed that data from twin studies include effects due to dominance and epistasis which would result in calculation of heritability estimates in the broad sense. Also, the fact that twins more often share the same environment than do nontwins may cause them to be more alike. Heritability estimates calculated from nontwin data normally include very little variation from other than additive genetic sources.

12.3.2 Apparent Heritability

Another method of measuring heritability is called *apparent* or *realized heritability*. In this procedure, the performance of the progeny of a group of selected individuals is compared to the performance of their parents. Actually, the average superiority in the performance of progeny is compared to the average superiority in the performance of their parents; it is a comparison of variations. The average increase in performance of progeny over the average of the population of which they are a part is often referred to as a *measure of genetic progress.* The average increase or superiority of the selected parents over the average of the whole population is called the *selection differential,* or *reach.* Figure 12.1 is a diagram illustrating selection differential. Note that the procedure requires the observance of at least two generations of individuals.

To illustrate, suppose that a group of selected dairy cows produces an average of 20,000 lb of milk per year. Assume that the average production of the larger population, possibly the breed average, is 16,000 lb. The cows selected are 4000 lb better than the average of the breed; this is the selection differential. If these superior cows are bred to bulls of equal genetic ability, further assume that the daughters of these matings ultimately produce at the rate of 17,400 lb of milk per year. Recognize that five or six years will pass during the process of collecting this information. The daughters pro-

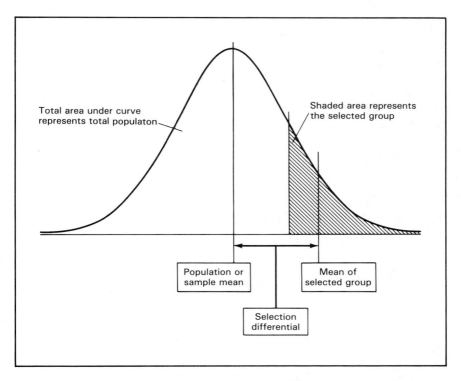

Figure 12.1   Diagrammatic definition of a selection differential.

duced at the rate of 1400 lb better than the breed average. This is a measure of that part of the superiority of the parents that was transmitted to the progeny, an indicator of the additive genetic variation. Since heritability is the ratio of additive genetic variation to total phenotypic variation, the estimate of heritability from this example is $1400 \div 4000 = 0.35$, or 35 percent.

Because the data on the parents and their progeny are not collected during the same year, it would tend to be difficult to account for the many environmental differences that might occur in different years. Using information from several sets of parent–progeny groups taken in different years would tend to increase the accuracy of heritability estimates calculated through this procedure. This is one reason why heritability estimates are quite often determined using statistical procedures comparing the performance of groups of relatives within the same year.

12.3.3 Using Correlation–Regression Analysis

The regression coefficient (b) is a measure of how much one variable is expected to change with each unit of change of another variable, on the average. If the assumption can be made that the variances of the two populations are not different, the regression of offspring on the average of both parents (midparent average) should be a reasonably valid measure of heritability in the narrow sense. Since an offspring received only $\frac{1}{2}$ its inheritance from any one parent, the regression of progeny on one parent measures only $\frac{1}{2}$ the genetic variation. Therefore, heritability can be estimated as 2 times the regression coefficient of sons or daughters on either their sires or dams.

Regression coefficients of one sib on another can also be determined. Since full sibs are expected to share approximately 50 percent of their genes, an estimate of heritability of any traits should be equal to 2 times the regression coefficient. Since half sibs share genes from only one common parent, their relationship is only 25 percent, so heritability can be estimated as 4 times the regression coefficient (refer to Table 12.2).

TABLE 12.2 Use of Regression and Correlation Coefficients
to Estimate Heritability

Source of Information	Regression	Correlation
Offspring and midparent	$h^2 = b$	$h^2 = r$
Offspring and one parent	$h^2 = 2b$	$h^2 = 2r$
Individual and full sib	$h^2 = 2b$	$h^2 = 2r$
Individual and half sib	$h^2 = 4b$	$h^2 = 4r$

As an illustration of how regression coefficients can be used to estimate heritabilities, refer again to the example in Chapter Eleven (Table 11.7) of the regression of birth weights of heifer calves on the birth weights of their dams. The regression coefficient in that example was 0.166. Since dams and daughters share 50 percent of their inheritance, the estimate of heritability from this example is 2 × 0.166 = 0.33.2, or about 33 percent.

If the variances of the two sample populations are not different, a correlation coefficient between two variables will approximately equal the regression of one of the variables on the other. Therefore, as seen in Table 12.2, correlation coefficients can be used to estimate heritabilities in the same manner as can regression coefficients. The correlation coefficient between identical twins should also be a measure of heritability, but for reasons that were presented earlier, it would probably be an estimate of heritability in the broad sense.

As a source of data, students in one of the author's classes were asked to estimate their heights and the heights of both natural parents. Over a period of several semesters, data were collected from 132 men and 76 women. The correlation coefficient between the heights of the male students and the average heights of their parents (midparent average) was 0.527. The coefficient of correlation between the heights of female students and the average heights of their parents was 0.551. Since these coefficients involve the midparent average, they represent estimates of heritability using the expression $h^2 = r$ from Table 12.2. From the same set of data, the correlation between heights of male students and their mothers was 0.316. Using the formula $h^2 = 2r$, the estimate of heritability is 0.632.

12.3.4 Using Analysis of Variance

Because members of the same family are genetically more alike than individuals from different families, an analysis of variance can be used to compare variation within families to variation among families. Such an analysis might involve full sibs or half sibs. Comparisons of variation within litters of hogs to variation among litters would involve full sibs. Half-sib comparisons could be made by comparing the variation within sire groups to variation among sire groups. This would be similar to comparing the variation within small one-bull herds of beef cattle to the variation among the herds. In properly conducted experiments to determine heritability estimates, attempts would be made to minimize the effects of other variables including differences in environment and breed differences.

The procedure involves performing an analysis of variance (ANOVA) among the family groups, whether full-sib or half-sib groups, and constructing an ANOVA table as explained in Chapter Eleven and illustrated again in Table 12.3. With respect to the use of analysis of variance to esti-

TABLE 12.3   Analysis of Variance Table

Source of Variation	Sums of Squares	Degrees of Freedom	Mean Squares
Among treatments (family or sire groups)	SS_A	$t - 1$	MS_A
Within treatments	SS_W	$n - t$	MS_W
Total	SS_T	$n - 1$	

mate heritabilities, the reference to treatments in the table refers to families, litters, or sire groups.

For comparisons involving full sibs, heritability can be calculated using the formula

$$h^2 = \frac{2MS_d}{MS_d + MS_W} \tag{7}$$

To solve for MS_d, we use

$$MS_d = \frac{MS_A - MS_W}{m} \tag{8}$$

where m is the average number of animals per family or sire group, and MS_A and MS_W are the mean squares among and mean square within, respectively, obtained from the ANOVA table. If the number of animals within any sire or family group is very much greater than the number of animals in other sire groups, the results can be overly influenced by the larger group. If there are unequal numbers of animals within groups, the following formula can be used:

$$m = \left(\frac{1}{t-1}\right)\left[n - \left(\frac{\Sigma m_i^2}{n}\right)\right] \tag{9}$$

where t is the number of sire or family groups, n is the total number of animals in the experiment, and the m_i's are the number of animals within each sire or family group. For comparisons involving half sibs, heritability can be estimated using the formula

$$h^2 = \frac{4MS_d}{MS_d + MS_W} \tag{10}$$

To illustrate the use of analysis of variance to estimate heritability, data from an unpublished study involving the performance of cutting horses will be used. The performance score for each horse was determined by three judges. The scores of 213 horses were separated into six groups by sires, and analyzed by analysis of variance. Table 12.4 includes some of the

TABLE 12.4   Results of a Study to Determine the Heritability
of Cutting Instinct in Quarterhorses

Sire Group	Horses per Sire Group	Sum of Scores	Mean Score
1	24	4911	204.6
2	30	6244	208.1
3	33	6986	211.7
4	34	7106	209.0
5	43	8948	208.1
6	49	9999	204.5
	213	44,194	207.5

TABLE 12.5   Results of Analysis of Variance of Data
Presented in Table 12.4

Source	Sum of Squares	Degrees of Freedom	Mean Squares
Among sires	1,457.9	5	291.6
Within sires	20,664.1	207	99.8
	22,122.0	212	

data obtained from the study. Table 12.5 presents the results of the analysis
of variance.

Before calculating the estimate of heritability, the value of MS_d [for-
mula (8)] must be determined. Since there are unequal numbers of animals
per sire group, applying formula (9) produces a value of m of 35.1. The
simple mean number of animals per sire group is 35.5, so the bias produced
by an unequal number of horses per sire group is probably not significant.
Now, calculating MS_d, we have

$$MS_d = \frac{291.6 - 99.8}{35.1} = 5.46$$

Since half sibs are involved, formula (10) will be used to calculate the
estimate of heritability:

$$h^2 = \frac{4 \times 5.46}{5.46 + 99.8} = \frac{21.84}{105.26} = 0.207$$

12.4 REPEATABILITY

Repeatability is a measure of the similarity between or among measures of
the same trait taken at different times in the life of the same animal. For
example, a statistical correlation between the first and second lactations of

a herd of dairy cows might be used as a measure of the repeatability of milk production. Since the genotype of a cow would not change between her first and second lactations, the total genetic effect on both lactations would be the same. This would suggest that the repeatability estimate for milk production should be at least as high as the heritability in the broad sense.

Another factor affecting the size of a repeatability estimate is the existence of permanent environmental conditions. For example, permanent udder damage would affect a cow's milk-producing ability for the rest of her life. Diseases and improper feeding practices that occur early in life may have permanent effects on the productive ability of an animal. These nongenetic factors would permanently influence the amount of variation among measures of a particular trait taken at different times in the life of an animal.

It should be evident that the size of a repeatability estimate for any trait is influenced greatly by the size of the heritability estimate for the same trait. The higher the heritability estimate, the higher the repeatability estimate. The range of repeatability estimates is from zero to 1.0, or 100 percent, the same as for heritability estimates. More will be presented in Chapter Thirteen about the applications and importance of repeatability.

STUDY QUESTIONS AND EXERCISES

12.1. Define each term.

apparent heritability	reach	selection differential
heritability	repeatability estimate	
heritability estimate	selection	

12.2. Explain how natural and artificial selection are similar.

12.3. Explain how natural and artificial selection are different.

12.4. Describe the genetic effect of selection.

12.5. Make a case for the futility of the argument that either heredity or environment is more important than the other.

12.6. Make a case for the position that heredity is more important than environment.

12.7. Make a case for the position that environment is more important than heredity.

12.8. Describe ways in which genetic variation can be subdivided.

12.9. List some of the factors that are included in environmental variation.

12.10. What is a genetic–environmental interaction?

12.11. How does a heritability estimate in the broad sense differ from one in the narrow sense?

12.12. Cite some examples of traits in farm animals whose heritability estimates are generally considered low.

12.13. Cite some examples of traits that are of medium heritability.

12.14. Cite some examples of traits that are highly heritable.

12.15. Why should the phenotypes of identical twins be more similar than those of dizygotic twins?

12.16. Assume that in a certain experiment, the heritability and repeatability of weaning weights of cattle were determined by correlation coefficients involving several hundred animals within the following three groups:
(a) Midparent and first calf, $r = 0.25$ (h^2 in the narrow sense)
(b) Identical twins, $r = 0.30$ (h^2 in the broad sense)
(c) First and second records of same cow, $r = 0.35$ (repeatability)
Explain why the values of r are different for each group.

12.17. What is the difference between heritability and repeatability?

PROBLEMS

12.1. Assume that the average weight at 154 days in a population of hogs is 170 lb and that of a group of boars and gilts kept for breeding purposes is 196 lb. (a) What is the selection differential? (b) If the heritability estimate for 154-day weight is 30 percent, how much progress can be expected in improving this trait in the next generation? (c) Given the condition already presented, how much would the next generation of pigs be expected to weigh on the average when they reach 154 days of age?

12.2. For some traits, making improvement may mean selecting for a decrease in value—backfat thickness in hogs is one example. Suppose that the average thickness of backfat in a small population is 38 mm. Gilts averaging 32 mm are selected for breeding purposes and will be bred to boars averaging 27 mm of backfat. (a) What is the average selection differential for boars and gilts combined? (b) If the heritability of backfat thickness is 45 percent, how much change in backfat thickness should be expected in one generation of selection? (c) What would be the average backfat thickness of the next generation?

12.3. Suppose that the average fleece weight in a certain breed of sheep is 7.8 lb and selected ewes and rams have an average adjusted fleece weight of 9.1 lb. (a) What is the selection differential for fleece weight? (b) If the next generation of sheep produce fleeces that average 8.3 lb, how much apparent progress has been made in improving this trait as measured in pounds? (c) From the information given, what is the apparent heritability of fleece weight?

12.4. Assume that the average weaning weight in a large population of beef cattle is 430 lb with a standard deviation of 40 lb. (a) What is the estimated statistical phenotypic variance of the population?(b) If the heritability estimate of this trait in the narrow sense is 30 percent, what is the additive genetic variance? (c) If it is known that the nonadditive genetic variance is 128 lb^2, what is the total genetic variance? (d) What is the estimate of heritability in the broad sense?

12.5. The average milk production for a certain population of dairy cows is 16,500 lb of milk per year, and the standard deviation is 1800 lb. **(a)** If the additive genetic variance is 907,200 lb^2, what is the estimate of heritability in the narrow sense? **(b)** If the epistatic variance is 4 percent and the dominance variance is 8 percent, what is the estimate of heritability in the broad sense? **(c)** If the selection differential is 0.8 of a standard deviation unit, how many pounds of milk would the next generation of heifers be expected to produce? **(d)** If the variance due to permanent environmental effects was known to be 150,000 lb^2, what is the repeatability estimate?

12.6. Following is the milkfat production record of a small herd of dairy cows:

Cow	First Lactation	Second Lactation	First Record of Half-Sister
1	590	550	610
2	600	610	560
3	550	560	540
4	560	580	520
5	620	600	570
6	580	580	530
7	550	600	590
8	540	560	550
9	540	540	570
10	570	550	540

Although these calculations would not normally be made from such limited amounts of data, calculate estimates of heritability using the first records of half sibs by **(a)** regression and by **(b)** correlation. **(c)** Calculate the repeatability estimate using the correlation between the first and second lactations.

12.7. Given the following set of data for four sire groups with regard to weaning score of calves, calculate the estimate of heritability using analysis of variance:

Sire A	Sire B	Sire C	Sire D
2	2	3	3
2	3	3	4
3	4	5	5
3	4	5	6
5	6	6	7
5	6	7	7
6	7	7	8
26	32	36	40

REFERENCES

BENDER, F. E., L. W. DOUGLASS, and A. KRAMER. 1982. *Statistical Methods for Food and Agriculture.* AVI Publishing Company, Westport, Conn.

BURNS, G. W. 1980. *The Science of Genetics,* 4th ed. Macmillan Publishing Co., Inc., New York.

FRANCOISE, J. J., D. W. FOGT, and J. C. NOLAN, JR. 1973. Heritabilities of and genetic and phenotypic correlations among some economically important traits of beef cattle. *Journal of Animal Science* 36 : 635.

LASLEY, J. F. 1987. *Genetics of Livestock Improvement,* 4th ed. Prentice-Hall, Inc., Englewood Cliffs, N.J. Chapters 7 and 8.

STANSFIELD, W. D. 1983. *Theory and Problems of Genetics,* 2nd ed. McGraw-Hill Book Company, New York.

WARWICK, E. J., and J. E. LEGATES. 1979. *Breeding and Improvement of Farm Animals,* 7th ed. McGraw-Hill Book Company, New York.

THIRTEEN

Improving Animals through Selection

By the very nature of genetic selection as a tool for animal improvement, it is a relatively slow process. Progress from selection can be realized in no less than the amount of time required to breed a group of animals and observe the performance in the next generation. By comparison, progress made through the improvement of the nutritional status of animals or by solving a disease problem can be realized in a few days. This certainly does not make the process of selection any more or no less important, but because of the time element, it is imperative that the process be well understood and carried out in a proper manner.

13.1 SINGLE-TRAIT SELECTION

A number of factors can affect the rate at which progress from selection can be realized. These factors apply to situations where selection is for a single trait or for several traits simultaneously. We will first describe these factors as they affect single-trait selection.

13.1.1 Size of the Selection Differential

As was presented earlier, genetic progress in improvement for any trait can be calculated by multiplying the heritability estimate by the selection differential ($H^2 \times$ S.D. = progress). It is obvious that if the size of the selection differential can be increased, a greater amount of progress can be realized.

To illustrate ways in which the size of the selection differential can be increased, consider a population of 10 heifer calves from which herd replacements will be kept. Although numbers are small for purposes of illustration, the same principles of selection would apply regardless of the size of the population. Assume that the weights of the 10 heifers ranked according to size were 500, 470, 452, 438, 425, 415, 402, 388, 370, and 340 lb. The mean of this sample is 420 lb. If the best two heifers were chosen, their mean would be 485 lb and the selection differential 65 lb (485 − 420 = 65). If the best three heifers were chosen, their mean would be 474 lb and the selection differential would be reduced to 54 lb. The selection differential for the best four heifers would be 45 lb, and for the best five heifers 37 lb. It should be clear that for a single trait, the fewer the number of animals chosen for breeding purposes, the greater the size of the selection differential can be, but this will result in the production of fewer numbers of offspring.

Reducing the size of the selection differential by selecting fewer replacements can be carried only so far in that a minimum number of animals must be retained to replace the poor-producing and nonproducing females in the herd. Depending on the quality of the herd and the amount of improvement desired, anywhere from 10 to 30 percent of the females in a commercial breeding herd should be replaced each year. To accomplish this in sheep and cattle herds, about one-fourth to two-thirds of the female progeny must be retained as possible replacements.

Because of the greater numbers of offspring produced in swine herds and poultry flocks, a smaller percentage of replacement females is needed to maintain population size. For this reason, selection differentials are generally larger in polytocous species than in monotocous species such as cattle, sheep, and horses. This represents another factor affecting the size of selection differentials; the greater the fertility rate, the fewer the number of females needed for replacement, and thus the larger the selection differential.

The necessity to keep fewer replacements obviously results in a greater number of progeny available for sale. For example, in commercial beef cattle herds, for every cow that fails to have a calf, two fewer calves are available for sale—the calf that the nonproducing cow did not have, and the heifer that must be kept in the herd to replace the nonproducing cow. This is based on the assumption that in commercial herds, nonproducing cows will not be retained in the breeding herd. The impact on gross sales is not as dramatic as implied, since the barren cow will usually return more than a weanling heifer. The point is, however, that the reduction in fertility does slow down selection progress as well as reduce the number of animals for sale.

Herd size is another factor affecting the potential size of the selection differential. Selection differentials can be larger in larger herds. This relates to a principle presented in Chapter Eleven that the larger the population, the greater the probability of having more extreme animals. In larger populations, there is a tendency to find values farther into the "tails" of the bell-shaped curve of normal distribution. Because of this, litter-bearing animals and poultry generally have an advantage over cattle, sheep, and horses in the rates of selection progress.

Selection differentials are also higher for males than for females. Since males can potentially produce so many more progeny, fewer are needed in the selection process, so their mean values can be higher for a given trait. This principle is enhanced with artificial insemination (AI). AI makes it possible to increase the numbers of progeny of any one sire manyfold, thus reducing the overall numbers of breeding males needed. The fewer the numbers of sires needed, the more superior they can be, and the greater the size of the selection differential can be. A similar principle can now be applied to females through the processes of superovulation and embryo transfer.

Data in Table 13.1 represent a numerical illustration of how changes in numbers of replacements needed can change the size of selection differentials. Selection intensity is defined as the ratio of selection differential to the standard deviation of the trait. If the standard deviation for weaning weight in a herd of beef cattle is 36 lb and the observed selection differential among the replacement heifers is 42 lb, the selection intensity is 1.17 units (42 ÷ 36). It is the proportion of the selection differential per unit of standard deviation. For the selection intensity to be 1.17, note in the table that

TABLE 13.1   Selection Intensity as Affected by the Proportion
of the Population Kept for Breeding

Percent of All Progeny Kept for Breeding	Selection Intensity (Selection Differential per Unit of Standard Deviation)
70	0.50
60	0.64
50	0.80
40	0.97
30	1.17
25	1.25
20	1.40
15	1.55
10	1.76
5	2.05
1	2.64

about 30 percent of the calf crop will need to be retained. Assuming a 1 : 1 sex ratio, this represents about 60 percent of the heifer calves. If only 10 percent of the calves were to be retained, the selection intensity would be 1.76, and the selection differential could be as high as 63 lb (36 × 1.76).

13.1.2 Size of the Heritability Estimate

Referring again to the formula for calculating genetic progress, H^2 × S.D. = progress, it can be seen that in addition to the size of the selection differential being a factor affecting the amount of progress expected, the size of the heritability estimate is also a factor. In this case, however, we are much more limited in our ability to control the size of the heritability estimate. Within certain limits, once it is determined which trait is to be improved, the size of the heritability estimate for that trait should be considered pretty much as a constant.

Recall that the mathematical expression for calculating a heritability estimate (in the narrow sense) is the ratio of the additive genetic variation to the total variation. Remember, too, that in general there are two major sources contributing to total variation, genetic and environmental. In experiments designed to determine heritability, the more that can be done to reduce environmental variation, the smaller the total phenotypic variation will be. If the genetic variation remains the same, it will represent a larger proportion of the total variation, resulting in a larger heritability estimate.

While it is recognized that within certain limits, if the amount of environmental variation can be reduced, the size of a heritability estimate can be increased, we should operate on the premise that once an estimate has been properly determined experimentally, it will remain relatively constant. Do not expect, for example, that through some drastic change in environmental conditions in a sow herd, the heritability of litter size can be changed from approximately 10 percent to something closer to 30 percent.

The following two illustrations will help demonstrate the effect of how the size of heritability estimates can affect the amount of progress in genetic improvement expected in one generation. As was just pointed out, litter size in swine has a heritability estimate of approximately 10 percent. Suppose that the average litter size at birth in certain breed of hogs is eight pigs. If the boars and gilts kept for breeding purposes were chosen from litters that averaged 10 in number, the selection differential is two pigs. The amount of progress expected in the next generation would be estimated as 10 percent of the selection differential, or about 0.2 pig. The actual average litter size expected from the selected sample would be 8.2 pigs, seemingly not much of an improvement over the average of 8.0-pig litter size of the parents.

The second example involves a trait with a higher heritability estimate that will result in greater genetic progress. Fleece weight as measured in pounds of wool per mature sheep has a heritability estimate of approximately 50 percent. If the average fleece weight in a certain breed of sheep is 8.0 lb, and the average of selected rams and ewes is 10 lb, the selection differential will be 2 lb. Note that the numbers so far except for the size of the heritability estimate are the same as those in the example above. The amount of genetic progress in this example is estimated to be 50 percent of the selection differential, or 1 lb of wool. The actual performance of the second generation of sheep should be about 9 lb of wool. The difference in the amount of improvement realized in these two examples is due to the differences in the sizes of the two heritability estimates.

The importance in knowing the estimates of heritability for traits we are trying to improve allows us to choose the proper breeding method to obtain that improvement. For traits with medium to high estimates of heritability, genetic selection, or mass selection, is an ideal method of seeking improvement. It will be shown later how outbreeding can be used to improve traits with lower estimates of heritability.

A knowledge of the size of the heritability estimate for a given trait provides some appreciation of an animal's potential breeding value for that trait. When the numerical value of the genetic progress expected from using a certain animal in a breeding program is added to the population or herd average for that trait, it provides an estimate of the breeding value of that animal. *Breeding value* is defined as the average level of performance expected among the progeny of a given animal or group of animals if bred to animals of equal quality. Much of the effort put forth in any selection program is aimed at trying to learn as much about potential breeding values as possible.

13.1.3 Generation Interval

The formula that has been presented for determining genetic progress, $H^2 \times$ S.D., is actually a measure of the progress that can be obtained in one generation. The average amount of progress that can be obtained within one year depends on how long is required to turn over one generation, this being the generation interval. *Generation interval* is defined as the average age of parents when all their offspring are born. It is the average length of time between the births of certain animals and the births of all their progeny. For example, if a cow has 10 calves in her lifetime when she was 2, 3, 5, 6, 8, 9, 10, 11, 12, and 14 years of age, respectively, the generation interval just for her family is 8 years. In reality, the generation interval for all cattle is about 6 or 7 years, one year for poultry, for hogs 2 to 3 years, and for humans about 30 to 35 years.

Generation interval (G.I.) affects genetic progress in that the longer the generation interval, the slower the overall progress. To estimate the expected genetic progress per year, the formula ($H^2 \times$ S.D.) ÷ G.I. can be used. If the heritability of a certain trait is 0.4, the selection differential 100 units, and the generation interval 5 years, the average progress expected in one year is 8 units, (0.4 × 100) ÷ 5. If the generation interval could be lowered to 4 years, the progress per year could be increased to 10 units. By the very biological nature of a species, greater progress can be obtained in those with shorter generation intervals. The fact that poultry and swine have shorter generation intervals than cattle, sheep, and horses, combined with the fact that they produce so many more progeny, accounts for most of why so much greater progress has apparently occurred in poultry and swine over the past 30 to 40 years.

Generation intervals can be shortened by keeping females in a herd for a shorter period of time. In the illustration used earlier, if the cow had been replaced after having her sixth calf, when she was 9 years old, the generation interval would have been shortened to 5.5 years. Economic factors can affect how much generation intervals can be shortened, especially in cattle and horses. The cost of growing a female to breeding age must be pro-rated over her productive years, so the longer she remains in the herd, the lower her annual cost. If the number of productive years is reduced too much, the returns may not cover the costs. There must be a balance between the greater progress obtained from a shorter generation interval, and the lower costs of a longer one.

13.2 SELECTION FOR MORE THAN ONE TRAIT

The effects of selection on genetic progress presented to this point have involved selection for one trait at a time. The same factors are involved when selecting for improvement in more than one trait at a time; however, the progress is not rapid. If the heritabilities are about the same for each trait, and the traits are inherited independently, the effectiveness of selection for any one trait will be less by approximately the square root of the number of traits involved. If selection is for two independent traits of equal heritability, selection for either one is $\sqrt{0.5}$, or about 71 percent as effective as selection for either trait alone. Notice, however, that selection effectiveness for one trait when two are selected together is more than 50 percent of that for selection for either trait alone. It is more efficient to select for two traits at a time than to select for a certain level of improvement in one followed by a certain level of improvement in the second.

13.2.1 Genetic Correlations

One factor that can influence the rate of progress when selection is for two or more traits is the genetic correlation between two of the traits.

A *genetic correlation* exists when genes affecting one trait also affect another trait. This phenomenon was described earlier as pleiotropy. Genetic correlations are apparent when pleiotropic effects of several genes work together on the same two traits.

A positive genetic correlation exists when selection in one trait not only results in improvement in that trait, but also results in improvement in a second trait. The higher the correlation, the higher the apparent genetic relationship between traits. In beef cattle, sheep, and hogs, animals that gain the fastest in the feedlot are usually among the most efficient in converting feed to gain. This means that selection for increased growth rate should result in some improvement in feed efficiency. However, since the correlation is not perfect (equal to 1.0), more progress can be made in improving feed efficiency by including it in the selection program. In the past, less attention has been given to direct selection for efficiency than for growth rate because efficiency requires more time and effort to measure.

If two traits are negatively correlated, selection for progress in one will result in a loss of progress in the second. Maximum progress cannot be obtained for both traits even though selection may occur for both. Because of the negative genetic correlation between percent butterfat and total milk yield in dairy cattle, breeders have not been able to develop a breed that will produce the volume of milk of the Holstein and the fat percentage of the Jersey.

For many traits, maybe most of them, no apparent genetic correlations exist, or they are so low as not to be significant. For example, there is no apparent relationship between weaning weights in sheep and the length of wool fibers. To obtain maximum improvement in either or both traits, selection would have to be practiced for both.

13.2.2 The Selection Index

The calculation of a *selection index* involves the separate determination of the values of two or more traits, and then adding the values to arrive at a measure of average total merit. It allows for relatively low values in one trait to be offset by larger values in another trait. Of the three methods of multiple-trait selection, it is the most efficient. The other two methods are the tandem method and the independent culling method.

The *tandem method* involves selection for progress one trait at a time. When a satisfactory level of performance is attained in one trait, the next trait is considered. This method is not generally recommended, because of the added time needed to achieve progress in several traits. It is also difficult to maintain a satisfactory level in one trait while trying to obtain progress in another. The efficiency of this method would depend upon the genetic relationships between and among traits.

The *independent culling method* involves selection for more than one trait at a time. It differs from a selection index in that minimum values are set for each trait. Any animal that expresses a value below the minimum is not chosen for breeding purposes regardless of how high the values may be for other traits under consideration. In other words, superiority in one trait does not compensate for relatively low values in another trait. It is not uncommon for principles of independent culling to be combined with those of a selection index.

A selection index usually takes the following general form:

$$I = b_1X_1 + b_2X_2 + ... + b_nX_n$$

where the X's represent the values of the traits in which improvement is being sought, and the b's are coefficients or adjustment factors that take into consideration such things as heritabilities, economic importance, and other factors.

Since the size of the heritability estimate of a trait affects the amount of progress that can be expected from selection, traits with different heritabilities require that different amounts of pressure be exerted to obtain equal amounts of progress. Similarly, increases in improvement in two traits may result in different economic returns. Therefore, the more economically important trait may call for a greater intensity of selection.

Differences in the sizes of means and standard deviations may make further adjustments necessary. Suppose that 154-day weight in hogs, with a mean and standard deviation of 190 ± 10 lb, and loin eye area with a mean and standard deviation of 5 ± 1 square inches are parts of a selection index. If the unadjusted values of the two traits are merely added to obtain an index, 154-day weight will be getting considerably more emphasis because of its greater standard deviation. A formula of $I = W + 10L$, where W is 154-day weight and L is loin eye area, would produce an index with both traits receiving equal emphasis. Further adjustments would be necessary if differences in economic value and heritabilities were considered. To illustrate further, suppose that a breeder decides that 154-day weight is twice as important economically as loin eye area. The index above could then be modified to double the emphasis on 154-day weight: $I = W + 5L$.

Another way to compensate for differences in relative variation among traits is to convert the data to the standardized form using the formula

$$X' = \frac{X - \bar{x}}{s}$$

where X' is the symbol for the standardized variable, X is the actual value for an individual animal, $\bar{x}$ is the mean value for the trait (the herd average), and s is the standard deviation. The general form of the selection index given earlier can be modified by using standardized variables rather than actual values:

$$I = b_1 X'_1 + b_2 X'_2 + \ldots + b_n X'_n$$

Standardized variables can be negative, as in the case of observations below the mean.

Table 13.2 illustrates how a selection index can be used to determine the total average merit of individuals. Note that in this hypothetical example, animal 3 has the highest index even though it does not rank first in any of the three traits in the index.

An attempt should be made to keep the number of traits in an index to a minimum. It is best to avoid including traits that are relatively unimportant economically and traits with extremely low heritabilities. On the other hand, if a trait is especially desirable, a larger increase in overall merit will result from including it in the index rather than trying to select for it independently.

13.3 SOURCES OF SELECTION INFORMATION

The amount of genetic improvement that can be obtained through a selection program will depend to a large extent upon how successful the breeder has been at choosing animals with the best combination of genes that will contribute to those traits identified for improvement. The degree of success will be revealed when observation is made of the performance of the individuals in the next generation. Several sources of information are available to help determine the probable breeding value of an individual or group of individuals. Some of that information may come directly from the individual, or from relatives of the individual. Relatives can be categorized into

TABLE 13.2   Indexes and Ranking of Seven Individuals for Three Hypothetical Traits

Animal	Standardized Variables			Index $(I = A + B + C)$	Rank by Index
	Trait A	Trait B	Trait C		
1	2.0	0.9	1.0	3.9	2
2	1.7	−0.6	1.7	2.8	4
3	1.4	1.4	1.3	4.1	1
4	1.1	2.1	−0.1	3.1	3
5	0.8	0.6	−0.4	1.0	6
6	−0.3	1.3	0.8	1.8	5
7	−0.4	0.2	0.3	0.1	7

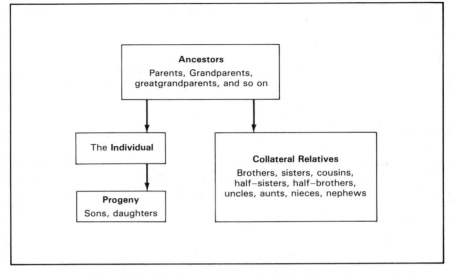

Figure 13.1   Relatives as sources of selection information.

three groups: ancestors, progeny, and collateral relatives. Figure 13.1 is an illustration of the definitions of the three categories of relatives.

13.3.1 Individuality

This source of information is based strictly on an individual's own phenotype and is often referred to as *mass selection.* The progress made from this type of selection depends to a large extent on the size of the heritability estimate of the trait being selected. The higher the estimate of heritability, the higher the correlation between the genotype and phenotype, and the greater the expected genetic progress. Much of the progress that has been made in livestock breeding has been due in large part to mass selection, especially for traits associated with body weight measurements and growth rates.

The merit or value of an individual must be determined by comparing the individual's phenotype to the average of the group in which it is selected. The comparisons should be made under conditions of controlled environment and excellent record keeping. The difference between the phenotype of an individual and the average of herdmates has been referred to previously as the selection differential. The *breeding value* (also called *probable breeding value, estimated breeding value,* and *predicted breeding value*) is determined by multiplying the selection differential by the heritability estimate for the trait and adding the herd average.

Suppose that in a herd of beef cattle the average yearling weight of bulls is 850 lb and the heritability estimate is 0.50. A bull with a yearling weight of 1000 lb would have an estimated breeding (BV) value of 925 lb:

$$BV = [(1000 - 850) \times 0.5] + 850 = 925$$

A breeding value of 925 suggests that if this bull were bred to cows genetically equal to him, his calves should weigh an average of 925 lb at a year of age.

Another way of expressing phenotypic values is in the form of a trait ratio. A *trait ratio* describes the performance of an animal relative to the herd average. In the previous example, the bull with the 1000-lb yearling weight has a yearling-weight ratio of (1000 ÷ 850)100, or 118. Ratios over 100 indicate values above the herd average. Ratios below 100 indicate values below the herd average. One advantage for the use of trait ratios is that comparisons can better be made among animals in different herds and in different years where environmental conditions are assumed to be different. Breeding values can also be calculated using trait ratios. For example, the 1000-lb bull with a breeding value of 925 has a yearling-weight ratio of 118 and a breeding value ratio of 109.

Many programs of genetic evaluation make use of transmitting abilities. In dairy bull evaluations, the term *predicted difference* (PD) or *sire comparison* is used. In the evaluation of beef bulls, *expected progeny difference* (EPD) is often used. The use of an estimated transmitting ability value attempts to show how much the average performance for a given trait will increase because a certain sire or dam is used.

Transmitting abilities are calculated as differences or deviations from average performance. In the example of the bull with a yearling weight of 1000 lb from a population that averaged 850, his superiority (selection differential) was 150 lb. The genetic portion of his superiority is indicated by the size of the heritability estimate (in this case 50 percent). Since only one-half the genetic superiority of an individual will be passed on in the gametes, the predicted transmitting ability of the bull would be calculated as (150 × 0.5)/2 = 37.5 lb. In other words, the adjusted yearling weights of his calves should average approximately 37.5 lb above the population average if he were bred to a randomly selected group of cows. The transmitting abilities of females can also be calculated using similar procedures. In reality, measures of transmitting ability are normally calculated from performance records of progeny rather than those of the parents.

Individuality is not as good an indicator of breeding values for traits with low heritabilities compared to traits with higher heritabilities. Consequently, progress from mass selection is expected to be relatively slow for traits with low heritabilities. This would suggest that other ways of seeking improvement should be used. Alternative methods of improvement would include outbreeding (linecrossing and crossbreeding) and better overall management practices related to better feeding and nutrition, disease control, and other factors of the environment. It is always wise to provide the best possible environment regardless of the breeding system used.

In deciding whether to keep an animal in the herd or to cull based on individual performance, it may be helpful to have information on more than one record for the individual. This is especially true for traits with medium to low heritabilities. Table 13.3 shows how the repeatability estimates increase for any trait as information from more records is obtained. The average of two or three or more records is a much better indicator of an animal's ability to perform than information from just one record. The impact of this statement is more pronounced for traits with single record repeatabilities in the range of 0.4 and below. The line in the table separating repeatability estimates above and below 0.5 is drawn to illustrate the impact of multiple records on repeatability.

The repeatability of weaning weight in beef cattle is about 40 percent. The average weaning weight of the first two calves provides a reasonably good estimate ($R = 0.57$) of what to expect of the dam in later years. The repeatability for three records is 0.67 compared to 0.57 for two. This is an increase of about 17 percent in reliability due to the third record. By the time a cow has had three calves, a breeder has a good idea of what to expect from the cow in the future. If the adjusted weights of the first two calves are particularly low, and certainly if the third calf is also a light one, the breeder can cull the cow and be reasonably confident that the cow was a poor producer.

The repeatability of litter size in swine is under 20 percent. To obtain information on a sow's litter size potential that is as reliable as two weaning weight records on a beef cow (0.57), five litters would have to be produced. In some swine herds, sows do not produce many more than five litters be-

TABLE 13.3   Effect of Increased Numbers of Records on the Size
of Repeatability Estimates

Number of Records	Repeatability of One Record					
	0.10	0.20	0.30	0.40	0.50	0.60
1	0.10	0.20	0.30	0.40	0.50	0.60
2	0.18	0.33	0.46	0.57	0.67	0.75
3	0.25	0.43	0.56	0.67	0.75	0.82
4	0.31	0.50	0.63	0.73	0.80	0.86
5	0.36	0.56	0.68	0.77	0.83	0.88
6	0.40	0.60	0.72	0.80	0.86	0.90
7	0.44	0.64	0.75	0.82	0.88	0.91

fore they are culled. By the time a breeder has enough information to be confident of a sow's ability, she may be ready to cull for other reasons. This is true for traits of low heritability, and thus low repeatability. This does not mean that mass selection is of no value in seeking improvement in traits with low heritabilities, but it does suggest that other methods may be of greater value.

Mass selection has some other limitations also. In some cases, a trait cannot be measured in but one sex. This is true with milk production in dairy cattle, egg production in poultry, and mothering instinct in all species. With some carcass traits measurements can only be made after an animal is slaughtered. With traits such as longevity in cattle and horses, time becomes a limiting factor in using individuality. With traits such as these, information about probable breeding values must be obtained from relatives.

13.3.2 Ancestors

A *pedigree* is a record of an individual's ancestors. The value of information about ancestors depends on how closely related an ancestor is. Individuals are related to each parent by a factor of 50 percent. *Relationship* refers to the percentage of genes that two animals have that are identical. For example, a sow receives one-half her genetic makeup from each parent; therefore she is related to her sire and dam each by one-half or 50 percent. One-half the genes an individual gets from one parent actually came from a grandparent, so the relationship of any animal to one grandparent is one-fourth or 25 percent. Inheritance is essentially a halving process—the relationship between an individual and an ancestor in any generation of the pedigree is one-half that of any member of the next succeeding generation. Figure 13.2 shows a format of a five-generation pedigree, indicating the average relationship of each ancestor to the individual whose pedigree it is (*X*). The relationship between an ancestor and any descendant, including progeny, is classified as direct relationship since genes that make them related were transmitted directly from one generation to the next.

Breeders should be careful not to put too much emphasis on certain outstanding ancestors in the pedigree. If, for example, a famous outstanding ancestor is in the sixth generation of the pedigree, the breeder must keep in mind that due to chance only a little more than 3 percent of that ancestor's genes will be present in the first-generation individual. Since no trait is 100 percent heritable, the contribution of the outstanding ancestor to the breeding value of the individual is even less. This points out that more emphasis should be placed on information from more recent ancestors. A parent carries twice the value of a grandparent in that regard. Another way to look at it is to note the number of individuals present in any generation of

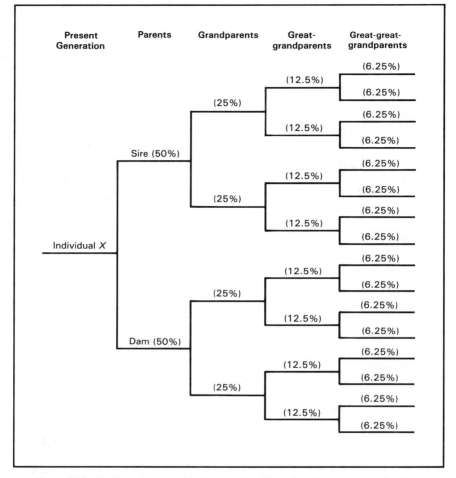

Figure 13.2 Pedigree bracket showing relationships of each ancestor to the first generation individual (X).

a pedigree. There are 32 ancestors in the sixth generation and each one contributes essentially the same number of genes as any other. On the average, the best individual will contribute the same as the poorest ancestor in that generation, $\frac{1}{32}$.

To be valuable in helping determine the breeding value of an individual, a pedigree must contain certain information about the performance of each ancestor, or at least the most recent ones. It should provide an idea of how each ancestor compared to its herdmates or contemporaries. In the past, most pedigrees did not provide this kind of information. More recently, however, this kind of information is becoming available in some breeds. On the other hand, even when performance data are available on the parents of an individual, it does not contribute as much to the accuracy of selection as does information on the individual being selected. For example, for a trait that is 30 percent heritable, information on both parents will increase accuracy of selection only about 12 percent above what individuality would contribute.

Pedigree information can be used to support what may already be known about an individual. For example, if performance records on a certain bull are very good and those of his parents are also favorable, this is valuable support for a conclusion that the bull may indeed be genetically superior. On the other hand, if the bull has a good performance record, but none of his ancestors do, it may be an indication that his superiority is due to heterozygosity or a "lucky combination of genes." This would be particularly significant if the bull's collateral relatives were also poor producers. The ultimate conclusion might be not to choose the only good bull in a bad family.

Information on members of the pedigree is useful in breeding programs for traits that are expressed in only one sex. Pedigree evaluation is important in the selection of dairy bulls for progeny testing, for example. A lifetime of records, or records from several years past maturity on parents, can be valuable in making early decisions about their progeny.

13.3.3 Collateral Relatives

Collateral relatives are those not related directly as ancestors or descendants, but indirectly by having obtained genes from one or more common ancestors. Examples of collateral relatives are full sibs, half sibs, cousins, aunts, uncles, nieces, and nephews. The most common types are full sibs and half sibs. Full sibs have the same two parents; half sibs have only one common parent. Pigs within the same litter are examples of full sibs, except for the possibility that a sow may have been bred to more than one boar. The calves in one small herd of cattle where all the cows were bred to the same bull are half sibs. The relationship between two full sibs is 50 percent, the same as between parents and progeny. The difference is that the common genes that sibs share came from the parents (collateral relationship), and the common genes shared by parents and progeny were actually transmitted from one to the other (direct relationship).

Boars on official test to determine their value for breeding purposes obviously cannot be slaughtered to obtain carcass data, for example. In these situations, littermate gilts and barrows (full sibs) can be slaughtered and their carcass information used as an indicator of the breeding value of the boar. Half-sib steers are often used to obtain similar data for breeding bulls on a performance test. Milk production records on a group of half sisters can be used to determine the breeding value of a dairy bull for milk production.

Although data on a full sib should be twice as valuable as data from a half sib because of the difference in their relationships, at least one precaution should be noted when using information from full sibs. Full sibs have the same dam as well as the same sire, so will have shared a similar maternal environment from conception to weaning. This is especially true in swine and poultry born or hatched during the same season. Adjustments can be made to compensate for the effects of shared environments.

13.3.4 Progeny

A *progeny test* involves a system whereby an estimate can be made of an animal's genotype or breeding value by observing the phenotypes of its progeny. Progeny testing is usually performed on males because they can produce so many more progeny than females. An individual contributes a sample one-half of its genes to any one offspring. The sample half received by one offspring will be different than the sample half received by another offspring—some genes will be the same, some will be different. However, all of an individual's genes should be equally represented within a group of progeny. Therefore, if a number of progeny are sired by a particular male when bred to a random sample of females, the average performance of the progeny is a reflection of the male's breeding value.

While progeny testing can be used to estimate breeding values for almost any trait, it is especially valuable for traits with low to medium heritability, such as reproductive efficiency and preweaning growth rate; terminal traits, such as carcass quality; and sex-limited traits, such as milk production. Generally, data on about five or six progeny are equal in value to that of an individual's own phenotype when individual data are available.

Time is one of the most limiting factors associated with progeny testing. By the time enough data have been collected on a bull or boar to determine his breeding value, considerable effort and expense have been devoted to the exercise. It is therefore important to select males for progeny testing that have shown promise from other sources of information, such as from parents and collateral relatives.

Based on the assumption that progeny receive one-half the dam's breeding value and one-half the sire's breeding value, a formula can be developed for calculating a sire index, or breeding value. If a bull is bred to a number of cows that represent a random sample of the population, the average of the progeny should be equal to one-half the sum of the average of all the cows and that of the sire:

$$\text{progeny} = (\text{sire} + \text{average of dams}) \div 2$$

Solving for the breeding value of the sire, or sire index, we obtain the following formula:

$$\text{sire index} = 2(\text{average of progeny}) - (\text{average of dams})$$

To illustrate the use of this formula, consider an example involving weaning weights in beef cattle. Assume that a certain bull is bred to a number of cows that represent the average in a very large herd. The average adjusted weaning weight of the sample was calculated to be 440 lb. The average weights of the calves at weaning age was 461 lb. The estimated breeding value of the bull is calculated to be $2(461) - (440) = 482$ lb.

The relationship between transmitting abilities and breeding values can illustrated from the example above. If breeding value in the illustration were expressed in terms of deviation from average performance, it would be 42 lb $(482 - 440)$. In those terms, the transmitting ability of the bull would be one-half his breeding value, or 21 lb. The performance of the calves would be the increase over the average of the population represented by the sum of the transmitting ability of the bull plus the average transmitting ability of the cows. In this example, since the cows represent the average of the population, their transmitting ability is zero. The average performance of the calves is 440 lb $+ 0 + 21$ lb $= 461$ lb.

STUDY QUESTIONS AND EXERCISES

13.1. Define each term.

ancestor	progeny
breeding value	progeny test
collateral relative	selection differential
estimated transmitting ability	selection index
generation interval	selection intensity
genetic correlation	standardized variable
independent culling	tandem selection
individuality	trait ratio
mass selection	
pedigree	

13.2. How does reducing the number of females kept for breeding purposes affect the size of the selection differential?

13.3. How does the number of offspring produced per female affect the percentage of total progeny that must be kept for replacement purposes?

13.4. Explain how size of the herd in which selection takes place affects the size of the selection differential.

13.5. Why are selection differentials generally higher in males than in females?

13.6. How can superovulation and embryo transfer increase potential genetic progress?

13.7. How does size of heritability estimates among several traits affect the relative progress that can be made from selection among those traits?

13.8. How may it be possible to increase experimentally the size of a heritability estimate for a given trait?

13.9. What is the relationship between generation intervals and the average annual genetic progress expected due to selection?

13.10. Compare generation intervals for several species of domestic animals (cattle, horses, swine, rabbits, chickens).

13.11. How can generation intervals be shortened?

13.12. What is the economic impact of shortening generation intervals?

13.13. When selection is being practiced for one trait, what effect will it have on another trait for which there is a positive genetic correlation?

13.14. Why can maximum progress not be obtained in the same animal for two traits between which there is a negative genetic correlation, even though selection pressure may be applied to both traits?

13.15. How do the independent culling and tandem methods of selection differ from the selection index method?

13.16. What is the relationship between individuality and size of heritability estimates with regard to progress that can be made from selection?

13.17. What are some advantages of using trait ratios rather than actual values in units such as pounds or inches?

13.18. What are some disadvantages or limitations of using individuality to predict breeding values?

13.19. Give some examples of how pedigree information can be useful in determining genetic potential of an individual.

13.20. List some possible limitations or shortcomings of using pedigrees as sources of selection information.

13.21. For what kinds of traits is information on collateral relatives most useful?

13.22. List some precautions to consider when using information from collateral relatives.

13.23. Why is progeny testing more often used to estimate breeding values of males than of females?

13.24. Give some examples of shortcomings of progeny testing as a method of predicting breeding values of individuals.

PROBLEMS

13.1. Following are the yearling weight ratios of 20 bulls on a performance test; the mean ratio is 100 :

102	102	98	101	117
109	104	103	112	111
103	110	97	113	96
115	101	102	106	105

(a) Assume that the top bull is to be used in an AI stud; what is his selection differential? (b) If the heritability of yearling weight is 48 percent, what is the estimated breeding value of the best bull? (c) What is the selection differential of the three best bulls? (d) What is the selection differential for the five best bulls? (e) What effect does increasing the number of bulls in the breeding pool have on the size of the selection differential?

13.2. In a certain flock of sheep the average fleece weight was calculated to be 7.7 lb. Following are the weights of the fleeces of the one-half of the ewe lambs that were above the average: 7.8, 7.9, 8.2, 8.2, 8.4, 8.5, 9.1, 9.2, 9.5, 9.8, 10.0, and 10.1. (a) If all 12 of these ewe lambs were kept for breeding purposes, what is the selection differential? (b) If only the top six ewes are kept for replacement in the breeding herd, what is the selection differential? (c) If the heritability of fleece weight is given as 40 percent, what is the average breeding value for the six best ewe lambs? (d) Calculate the average breeding value for the 12 ewe lambs. (e) If the 12 ewes were bred to a ram with an estimated breeding value of 8.6 lb, predict the average fleece weight expected in the progeny.

13.3. The heritability of weaning weights in lambs is about 30 percent. In a certain flock, the average weaning weight is 50 lb, the average of ewes selected for breeding purposes is 62 lb, and the generation interval is three years. (a) Calculate the selection differential. (b) If the ram's weaning weight were 62 lb, calculate the expected average weaning weight of the lambs in the next generation. (c) Calculate the expected average annual progress.

13.4. Listed below are the values of three different traits for 10 sows being considered for the breeding herd.

Sow	Weight at 154 days, W (lb)	Backfat Thickness, B (cm)	Average Weight of Litter Mates at Weaning, L (lb)
1	180	22	31
2	192	28	34
3	186	29	28
4	192	25	38
5	198	32	35
6	181	30	37
7	188	26	34
8	175	30	33
9	184	27	40
10	190	31	29

(a) Using the index $I = W + 2L - B$, calculate the indexes of the five best sows. (b) If the average herd index is 212, what is the selection differential for the five top sows? (c) If the heritability estimate of the index is 55 percent, what average index should exist among the progeny of the five top sows?

13.5. Listed below are individual observations, means, and standard deviations of three traits for six bulls being evaluated for a breeding sale.

Bull	Yearling Weight, W (lb)	Feedlot Rate of Gain, G (lb/day)	Loin eye Area, L (in.²)
1	880	3.5	8.1
2	800	2.7	8.7
3	980	3.3	7.3
4	850	2.9	8.0
5	875	3.1	8.3
6	935	2.5	7.9
Mean	805	2.7	7.5
Std. dev.	70	0.4	0.5

(a) Calculate the standardized variables for each trait for each bull. (b) Using the index $I = 10 + 3W' + 2G' + L'$, where W', G', and L' are the standardized forms of yearling weight, rate of gain, and loin eye, respectively, rank the bulls according to their indexes.

13.6. A certain bull is being considered for breeding purposes; his yearling weight was 980 lb. The heritability of yearling weight is estimated at 54 percent. (a) If the average yearling weight for the herd is 820 lb, what is the estimated breeding value of the bull? (b) Assuming that the calculated breeding value is an accurate measure of the bull's true breeding value, predict the average weaning weight of the calves sired by this bull if bred to average cows in the herd. (c) If the bull in part (a) is bred to a group of cows with yearling weights of 910 lb (adjusted to a bull basis), calculate the expected average yearling weights of bull calves from these cows.

13.7. Listed below are the first lactation milk production records of nine randomly selected cows involved in a progeny test. The records of nine half-sib daughters are also listed (all have the same sire). Calculate the estimated breeding value of the sire.

Cow	Cows' Records	Daughters' Records
1	13,500	17,450
2	12,200	16,700
3	12,600	15,900
4	14,700	16,800
5	11,900	16,800
6	15,300	15,200
7	12,100	15,750
8	12,900	17,000
9	13,800	14,900
Means	13,222	16,411

REFERENCES

JOHANSSON, I., and J. RENDELL. 1968. *Genetics and Animal Breeding.* W.H. Freeman and Company, Publishers, San Francisco.

LASLEY, J. F. 1987. *Genetics of Livestock Improvement,* 4th ed. Prentice-Hall, Inc., Englewood Cliffs, N.J. Chapters 8 and 9.

TERRILL, C. E. 1958. Fifty years of progress in sheep breeding. *Journal of Animal Science* 17 : 944.

WARWICK, E. J., and J. E. LEGATES. 1979. *Breeding and Improvement of Farm Animals,* 7th ed. McGraw-Hill Book Company, New York.

FOURTEEN

Relationship and Inbreeding

In addition to increasing the frequencies of favorable genes through selection, an animal breeder can have some control over the genetic makeup of animals by choosing a particular mating system. Mating animals that are more closely related or more alike than the average of the population will result in genotypes in the progeny that are more homozygous than their parents. This is called *inbreeding*. On the other hand, mating animals that are less alike or less closely related than the average of the population will result in progeny that are more heterozygous than their parents. This is *outbreeding*. Outbreeding is discussed in detail in Chapter Fifteen. Since genetic relationship is the key to whether a breeder is practicing inbreeding or outbreeding, some time and space should be devoted to a discussion of the principles and concepts of relationship.

14.1 RELATIONSHIP

Relationship refers to the proportion of identical genes that two animals have because they are members of the same family. The number of genes that any two animals have in common due to chance is generally not considered in calculating genetic relationship. Two animals may be related because one of them contributed genes to the other by way of their gametes, such as between parents and progeny. This is called *direct relationship*. Two animals may be related because they received identical genes from the same parent, grandparent, or other common ancestor. This is called *collateral relationship.*

14.1.1 Direct Relationship

Direct relationships exist between individuals and ancestors and between individuals and descendents. An individual is related to a parent because one-half that individual's genes were obtained from the parent. Since 50 percent of their genes are the same, their genetic relationship is 50 percent. This is called the *coefficient of relationship.* It can be expressed as a percentage or as a decimal fraction.

Referring to the hypothetical pedigree in Figure 14.1, Rosey is related to each member of her pedigree because genes were transmitted in a direct line from each ancestor by way of the gametes. Inheritance is essentially a halving process in that one-half the genes that Rosey obtained from her sire came from her grandsire Victor. One-half of his genes came from his sire and so on. On the average, each of the four grandparents in the pedigree is the source of one-fourth of the genetic makeup of Rosey. Rosey is therefore related to each grandparent by 25 percent. By the same principle, the eight great-grandparents in the pedigree are approximately equal contributors to the genotype of Rosey, and are related to her by 12.5 percent. Due to chance and the sampling nature of inheritance, it is understood that the percentages could vary more or less from the figures stated.

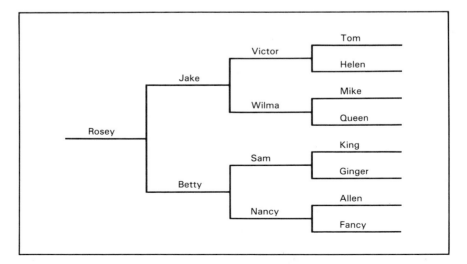

Figure 14.1 Hypothetical pedigree of Rosey.

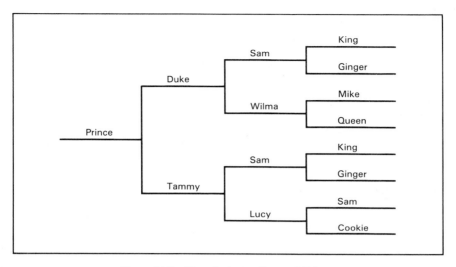

Figure 14.2   Hypothetical pedigree of Prince.

If an ancestor appears in an individual's pedigree more than once, the relationship between them is greater than if the ancestor had appeared only once. Consider the hypothetical pedigree of Prince in Figure 14.2. The ancestor Sam appears in the pedigree three times, twice in the third generation and once in the fourth. Recall that each of the four ancestors in the third generation of a pedigree contributes approximately 25 percent to the genotype of the individual in the first generation. Since Sam appears twice in this generation, he would be expected to be the source of 50 percent of the genotype of Prince. Further, because Sam appears once in the fourth generation, he is the source of another 12.5 percent of Prince's genetic makeup. The coefficient of relationship between Prince and Sam therefore amounts to the sum of the genetic contributions that Sam makes due to each appearance in the pedigree. The total relationship in this example is 0.625, the sum of 0.25 + 0.25 + 0.125. This is greater than the usual relationship between a sire and his son.

14.1.2 Collateral Relationship

The type of relationship that exists between two individuals due to the genetic contribution of a common ancestor is called collateral relationship. Collateral relatives are not directly related since they are not ancestors or descendants of each other. Consider again the pedigree of Prince in Figure 14.2. Duke and Tammy, the parents of Prince, are collaterally related because of their common sire Sam. Having one common parent would make them half sibs and would cause them to be related by about 25 percent on the average. Full sibs would have both common parents and would be related by 50 percent. Duke and Tammy, however, are actually related by more than 25 percent because Sam is also Tammy's maternal grandsire. The calculation of coefficients of relationship for collateral relatives such as this are a bit more complicated than for direct relatives. The procedure is discussed later in this chapter.

Several other examples of collateral relationship can be seen between members of the two pedigrees due to the presence of two common ancestors, Sam and Wilma. Rosey, in the first pedigree, is related to Prince, Duke, Tammy, and Lucy in the second pedigree. Prince, in the second pedigree, is related to Jake and Betty in the first pedigree. Notice that collateral relatives need not be in the same generation; they merely need to have at least one ancestor that appears in both pedigrees.

The definitions of full sibs and half sibs have been presented. Several other terms are used to describe certain collateral relationships in human families, and sometimes in domestic animals. The hypothetical family in Figure 14.3 will be used to illustrate. In the first generation, A, B, and C are not genetically related. A was the first husband of B, and C was her second husband. In the second generation, D and E are full sibs since they have both the same parents. F is a half sib to both D and E because they

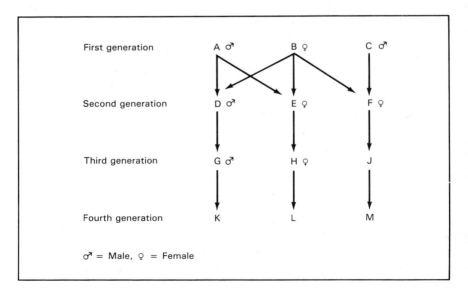

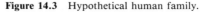

Figure 14.3 Hypothetical human family.

have only B as a common parent. H is a niece to D because she is the daughter of his full sib. Conversely, D is an uncle to H.

The children of full sibs, G and H, for example, are first cousins. The children of first cousins, K and L in the fourth generation, are second cousins. G and L are first cousins once removed, as are H and K. The children of half sibs, E and F, for example, can be called half cousins. Table 14.1 lists many of the standard relationships, both collateral and direct, and their coefficients.

14.1.3 Calculating Relationship Coefficients

The relationship between half sibs is 25 percent. Refer again to Figure 14.3 and the half sisters E and F. Let us assume for the purpose of illustration that their common mother B carries 10,000 genes in the nucleus of each body cell. A sample one-half of the total genes would be carried in each gamete, so each daughter would carry 5000 genes from her mother. Some

TABLE 14.1   Coefficients of Relationship between an Individual and Various Family Members

Family Member	Coefficient of Relationship	
	Fraction	Percent
Identical twin	1	100
Fraternal twin (full sib)	$\frac{1}{2}$	50
Full brother or sister, not a twin	$\frac{1}{2}$	50
Father or mother (sire or dam)	$\frac{1}{2}$	50
Son or daughter (progeny)	$\frac{1}{2}$	50
Half brother or sister (half sib)	$\frac{1}{4}$	25
Grandparent	$\frac{1}{4}$	25
Grandchild	$\frac{1}{4}$	25
Uncle or aunt	$\frac{1}{4}$	25
Niece or nephew	$\frac{1}{4}$	25
Double first cousin (four common grandparents)	$\frac{1}{4}$	25
First cousin (two common grandparents)	$\frac{1}{8}$	12.5
Great grandparent	$\frac{1}{8}$	12.5
Half cousin (one common grandparent)	$\frac{1}{16}$	6.25
First cousin once removed	$\frac{1}{16}$	6.25
Great-great-grandparent	$\frac{1}{16}$	6.25
Second cousin	$\frac{1}{32}$	3.125
Second cousin once removed	$\frac{1}{64}$	1.56
Third cousin	$\frac{1}{128}$	0.78

of the 5000 genes that daughter E obtained from her mother would be the same ones that daughter F received; some would be different. Due to chance it would be expected that one-half the genes received by daughter E would have also been received by F; therefore, approximately 2500 of the 10,000 genes in their body cells should be the same. The genetic relationship between half sisters E and F is thus 25 percent.

Because individuals D and E have the same mother, they also share approximately 25 percent of their genes. In this case, however, D and E are full sibs since they also have the same father. Because of this, another 25 percent of their genes should be identical. The total percentage of genes they have in common is equal to the sum of those they received from both parents. Thus the coefficient of relationship between full sibs is 50 percent.

One method of calculating the genetic relationship between any two collateral relatives involves a system of arrow diagrams and path coefficients. This can be illustrated by calculating the coefficient of relationship between Prince and Rosey in Figures 14.1 and 14.2. Begin by replacing the names of individuals in the pedigrees with letters or numbers for simplification. Be sure that any individual that appears more than once gets the same letter or number each time. It is helpful to orient the simplified pedigrees so that one is directly above the other, as illustrated in Figure 14.4. Locate the ancestors that are common to both pedigrees, drawing a box or circle around them to set them apart. W and S are common ancestors.

In preparing an arrow diagram as shown in the lower part of Figure 14.4, each individual in the two pedigrees is listed only once in the arrow diagram although they may appear in the pedigrees several times. By drawing arrows from each common ancestor to each succeeding descendent, and finally to Prince and Rosey, it can be seen that individuals that appear more than once will have more than one arrow leading from them. For example, S appears four times in the two pedigrees and has four arrows leading away from him. Notice that the direction of the arrows is always away from the common ancestors and toward the descendents, in other words, in the direction in which genes are transmitted.

A question can be raised as to why individuals K and G (parents of S) and M and Q (parents of W) are not listed in the arrow diagram. These four individuals appear only as the parents of S and W, so their genes are already accounted for. If one of them had also appeared anywhere other than as the parents of S or W, they would have had to be listed in the arrow diagram.

The next major step is to prepare a table of all possible pathways connecting one collateral relative, P in this case, to the second, R, through each common ancestor, including all individuals in between. This process is illustrated in Table 14.2. Again, highlight each common ancestor with a

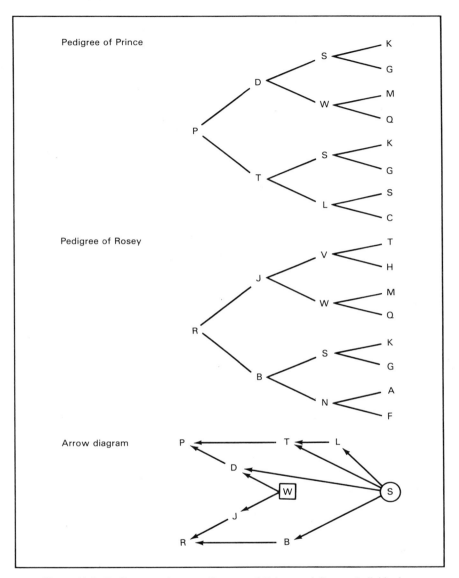

Figure 14.4 Pedigrees and arrow diagram of Prince and Rosey. Individuals are identified by the first letters of their names.

TABLE 14.2 Pathways and Path Coefficients for Calculating
the Coefficient of Relationship between Prince and Rosey

Pathway	Number of Arrows, n	$\frac{1}{2}^n$
P ⟵ D ⟵ W⟶ J ⟶ R	4	0.06250
P ⟵ D ⟵ S ⟶ B ⟶ R	4	0.06250
P ⟵ T ⟵ S ⟶ B ⟶ R	4	0.06250
P ⟵ T ⟵ L ⟵ S ⟶ B ⟶ R	5	0.03125
	$R = \Sigma (\frac{1}{2})^n$	0.21875

box or circle. Notice that the arrows always point toward the collateral relatives, and away from the common ancestors.

The formula for calculating relationship is $R = \Sigma (\frac{1}{2})^n$, where R is the coefficient of relationship; Σ, the uppercase Greek letter sigma, means to add; and n is the number of arrows in each separate pathway. For each pathway, raise $\frac{1}{2}$ to the power of the number of arrows in that pathway. The sum of all path coefficients is the unadjusted relationship coefficient for P and R. They are related by about 22 percent, almost as much as half sibs. Later, it will be shown how inbreeding in a common ancestor or in one or both of the collateral relatives affects the coefficient.

14.2 PRINCIPLES OF INBREEDING

A considerable amount of space has just been given to a discussion of genetic relationship and the calculation of coefficients of relationship. Having an appreciation for and an understanding of genetic relationship should now make it easier to understand inbreeding and its application as a breeding tool. Inbreeding is usually defined as the mating of animals that are more closely related than the average of the population. If parents are related, their progeny are inbred. The more closely related the parents, the more highly inbred the progeny. More will be presented later about degrees of inbreeding and calculation of inbreeding coefficients.

14.2.1 Genetic Effect of Inbreeding

The only genetic effect of inbreeding is that it increases the number of pairs of genes that exist in the homozygous state. The percentage of loci with different alleles decreases in favor of pairs of genes that are the same. This process can be well illustrated using an example of self-pollination in plants. This is the most extreme form of inbreeding that can be practiced.

TABLE 14.3 Illustration of Self-Pollination Showing the Genetic Effect of Inbreeding

Generation	Genotypes and Numbers of Each			Percent of Heterozygous Gene Pairs	Percent of Homozygous Gene Pairs	Frequency of Gene A (%)
	AA	Aa	aa			
0	0	128	0	100	0	50
1	32	64	32	50	50	50
2	32 + 16	32	16 + 32	25	75	50
3	48 + 8	16	8 + 48	12.5	87.5	50
4	56 + 4	8	4 + 56	6.25	93.75	50
5	60 + 2	4	2 + 60	3.125	96.875	50
6	62 + 1	2	1 + 62	1.56	98.44	50

An illustration of the genetic effects of inbreeding is shown in Table 14.3. The zero or parental generation is assumed to all be heterozygous. The same number of individuals is maintained in each generation to simplify calculations. As inbreeding proceeds, the percentage of homozygous pairs of genes increases and the percentage of heterozygous pairs decreases. The relative frequencies of the two alleles remains the same from generation to generation. The frequencies of the genotypes changes in favor of the homozygotes. Suppose that one of the homozygotes were being favored and the other suppressed by environmental factors as inbreeding progressed; the frequencies of the two alleles would be changed. However, the change would be caused by selection and not inbreeding.

The rate at which homozygosity increases due to inbreeding in any individual depends to a large extent on the closeness of relationship between the parents. In the example of self-pollination, the rate is quite high. In a series of full-sib matings, the rate of change in homozygosity from one generation to the next would be about one-half that of self-pollination. A series of half-sib matings would result in an increase in homozygosity at about one-half the rate of that of full-sib matings. In animals the rates of increases in homozygosity from inbreeding are generally slower than in plants, but the trend is the same.

14.2.2 Linebreeding

The only criterion for inbreeding to occur is that the two individuals being mated are related by more than the average of the population. The relationship may be direct, or it may be due to one or more common ancestors. As long as the animals being mated are related, the progeny will be inbred.

Linebreeding is a kind of inbreeding where the relatives being mated are chosen because of a particular common ancestor. The goal is usually to incorporate into the linebred progeny as many genes as possible from a certain popular or superior ancestor. In the process, it may also involve the selection of better-quality animals for other traits. The genetic effect, however, is the same as for any plan of inbreeding; it increases homozygosity.

Figure 14.5 contains a pedigree and arrow diagram illustrating ordinary inbreeding. It involves the relationship between the parents of individual X due to the presence of four unrelated common ancestors. Figure 14.6 illustrates linebreeding. The parents of individual Y are related due to one common ancestor that appears several times. The relationship between the

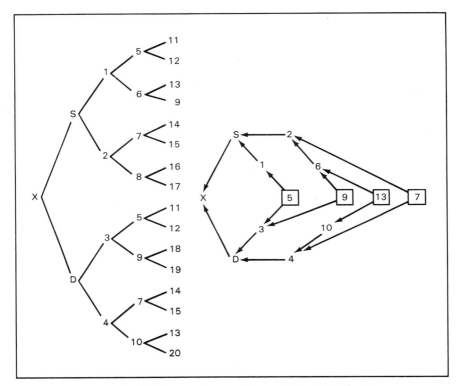

Figure 14.5   Pedigree and arrow diagram illustrating inbreeding.

parents in X's pedigree is about the same as between the parents in Y's pedigree, so the degree of inbreeding is about the same in the two pedigrees.

14.3 PHENOTYPIC EFFECTS OF INBREEDING

Inbreeding is often associated with a decline in performance and vigor and an occasional appearance of genetic defects in animals. Most animal breeders are aware of this and may try to avoid the use of inbreeding as much as possible. The belief that inbreeding is generally bad is so strong in humans that marriages among members of the same family, including first-cousin marriages, are usually avoided, and in some states are prohibited by law. Where the effects and uses of inbreeding among domestic animals are well understood, considerable benefits can be obtained from the practice.

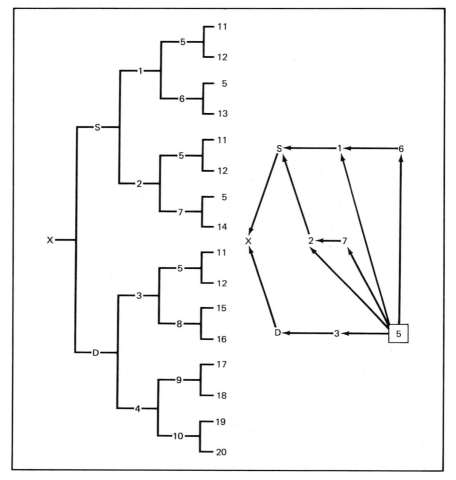

Figure 14.6   Pedigree and arrow diagram illustrating linebreeding.

14.3.1 Physiological Explanation

Inbred animals are generally less able to adjust to their environment compared to outbred animals. There is usually a reduction in fertility rates, increases in deaths and susceptibility to disease, and slower growth rates. The negative effects of inbreeding are referred to as *inbreeding depression*. Many of the adverse effects of inbreeding are known to be caused by recessive genes. These adverse effects are probably not due to the presence of any particular pair of recessive genes, but to the combined effects of many pairs.

Since inbreeding increases the incidence of homozygosity, more pairs of recessive genes are expected to occur among inbred animals. Their frequency is not increased in the population; they merely appear more often as homozygotes, whereas before inbreeding more occurred as heterozygotes. One explanation of why homozygous recessive genotypes result in detrimental effects is because of deficiencies of important or essential enzymes that would have been produced by their dominant alleles. Recessive genes may sometimes result in the production of abnormal or defective proteins or other compounds. Imbalances or deficiencies of certain hormones may also be associated with recessive genotypes.

As will be explained in greater detail later, outbreeding results in an increase in heterozygosity of genes. The phenotypic results are usually more positive and are referred to as *hybrid vigor* or *heterosis*. Hybrid vigor may express itself in the form of increased fertility, reduction in susceptibility to disease, and increases in growth rate. To the extent that inbreeding reduces heterozygosity, it tends to reduce the positive effects of hybrid vigor and thus magnifies inbreeding depression.

14.3.2 Effect of Heritability on Inbreeding Depression

The greatest amount of inbreeding depression is observed among those traits controlled to a large extent by nonadditive gene action. Examples of nonadditive gene action include dominance and recessiveness, epistasis, and overdominance. Additive genes seem to be affected least by increases in homozygosity. Traits affected the most by additive genes and least by nonadditive genes are the traits with the highest levels of heritabilities. Recall that a heritability estimate (in the narrow sense) is a measure of the proportion of the total variation in a trait controlled by additive gene action. Traits controlled mostly by nonadditive gene action typically have the lowest heritabilities, particularly in the narrow sense. It follows that the amount of inbreeding depression observed among traits is inversely related to the sizes of the heritability estimates for those traits.

14.3.3 Practical Uses of Inbreeding

Inbreeding can be used along with selection to uncover and eliminate detrimental recessive genes. Not only can recessive homozygotes be eliminated, but their heterozygous parents can also be disposed of, at least for breeding purposes. The rate at which recessive genes can be uncovered depends upon the closeness of inbreeding.

Inbreeding can also increase the homozygosity of favorable genes. Along with selection, the best-performing animals among an inbred population can be retained for breeding purposes. Inbred animals that tend to exhibit superior performance tend to be more homozygous for the genes that control that superiority. When they are used for breeding, they will be more likely to pass that superiority along to their progeny, because their genotypes are more homozygous. The phenomenon whereby animals are able successfully to transmit their own characteristics to their progeny is called *prepotency*. Inbred animals tend to be more prepotent.

A very practical use of inbreeding is to produce inbred lines to be used in outcrossing programs. Superior males that were the result of several generations of inbreeding and selection should produce excellent results in both purebred and commercial breeding programs. When bred to relatively unrelated females in purebred herds, benefits should be obtained from heterosis as well as from prepotency of the superior inbred sires. In commercial herds, it seems especially practical to breed superior inbred bulls and boars to crossbred females to maintain a three-breed rotational crossbreeding program.

Another practical use of inbreeding is in the form of linebreeding. As defined earlier, linebreeding is a kind of inbreeding where an attempt is made to concentrate the genes of one superior ancestor into the herd. Because of the efforts put forth to choose animals with a particular ancestor, it is possible also to attempt to select for better-quality animals. Although there are exceptions, linebreeding in many herds may not be as intense as is ordinary inbreeding.

In a discussion of the practical uses of inbreeding, at least two conclusions can be made. One is that the most benefit can be obtained from inbreeding when it is used in conjunction with selection. Inbreeding just for the sake of inbreeding, except for purposes of research, does not appear to be practical. The second conclusion is that inbreeding is to be used for the production of seedstock, especially males. This means that the practice of inbreeding will generally be limited to the larger purebred herds. Commercial herds can benefit from the use of inbred sires produced by purebred breeders. It is not practical to use inbred females in commercial herds because of their inferior performance compared to outbred females. More is presented concerning this principle in Chapter Fifteen.

14.4 INBREEDING COEFFICIENTS

An *inbreeding coefficient* provides a measure of the increase in homozygosity in an individual over that of the parents due to inbreeding. It is also a measure of the decrease in heterozygosity compared to that of the parents or some base population. In a very general way, inbreeding coefficients are used as indicators of homozygosity in inbred animals. They can also be used as indicators of the degree to which the parents of inbred animals are related. For these reasons, it is important that animal breeders understand and be able to calculate inbreeding coefficients.

14.4.1 Calculation of Inbreeding Coefficients

Generally, the inbreeding coefficient of an individual is one-half the relationship between the parents of that individual. So the calculation of an inbreeding coefficient is a matter of calculating the relationship coefficient between the parents and dividing by 2. This is true if neither of the parents or the common ancestors are inbred. The formula for calculating an inbreeding coefficient in this case is $F = \Sigma \left(\frac{1}{2}\right)^n/2$, where F is the symbol for the inbreeding coefficient, and the numerator of the formula is identical to the formula for calculating relationship coefficients.

Half sibs are related by 25 percent. The offspring of a half-sib mating would have an inbreeding coefficient of 12.5 percent. Following is a pedigree and arrow diagram of an individual produced from a half-sib mating:

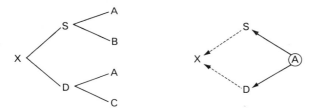

The pedigree involves one pathway from S to D through A, and includes only two arrows: S←Ⓐ→D. The relationship between S and D is equal to $\left(\frac{1}{2}\right)^2$ or $\frac{1}{4}$ or 25 percent; the inbreeding coefficient of X is one-half that. When calculating inbreeding coefficients, be careful not to count the arrows from the parents to the progeny. If it can be remembered that the procedure for calculating inbreeding coefficients is essentially the same as for calculating relationship coefficients, this error can be avoided.

Following is a pedigree and arrow diagram of a full-sib mating:

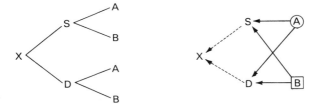

Full sibs have two common parents, so two pathways can be drawn from S to D through them: $S \leftarrow \boxed{A} \rightarrow D$ and $S \leftarrow \boxed{B} \rightarrow D$, with two arrows each. The relationship between S and D is the sum of the two path coefficients $(\frac{1}{2})^2 + (\frac{1}{2})^2$, or $\frac{1}{2}$ or 50 percent. It has already been established that full sibs are related by 50 percent; this shows how it can be calculated. The inbreeding coefficient of the offspring of a full-sib mating is one-half the relationship between full sibs, or 25 percent. Other pedigrees may be more complicated, but the principle for calculating inbreeding coefficients remains the same.

14.4.2 Effect of Inbreeding in a Common Ancestor

Inbreeding coefficients and relationship coefficients are both affected by inbreeding that may exist in a common ancestor. The basic formula for calculating an inbreeding coefficient must be modified if a common ancestor is inbred. The adjusted formula becomes $F = \frac{1}{2} \Sigma (\frac{1}{2})^n (1 + F_a)$, where F_a is the inbreeding coefficient of the common ancestor. Figure 14.7 includes a pedigree and arrow diagram where the common ancestor of the sire and dam is inbred.

Without drawing the pathways and counting arrows it can be seen that the sire (S) and dam (D) are half sibs; they have one common parent (B). The relationship between S and D would initially be expected to be 25 percent, and the inbreeding coefficient of X would be expected to be 12.5 percent. However, the common ancestor (parent in this example) is the product of a full-sib mating. Because E and F are full sibs, they are related

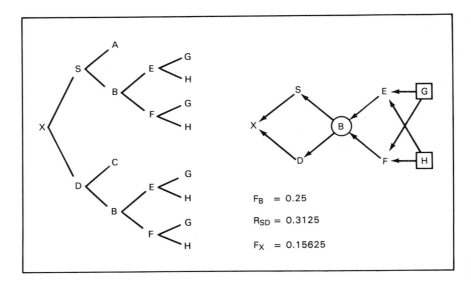

$F_B = 0.25$

$R_{SD} = 0.3125$

$F_X = 0.15625$

Figure 14.7 Pedigree with an inbred common ancestor.

by 50 percent and the inbreeding coefficient of B is 25 percent. The inbreeding coefficient of X must be increased by the amount of the inbreeding in the common ancestor. The adjusted inbreeding coefficients is $F = 0.125$ $(1 + F_B) = 0.125(1.125) = 0.15625$ or about 15.6 percent.

14.4.3 Effect of Inbreeding on Relationship Coefficients

As explained previously, the inbreeding in the common ancestor in Figure 14.7 increases the relationship between S and D. Instead of their coefficient of relationship being 0.25 because they are half sibs, it is actually

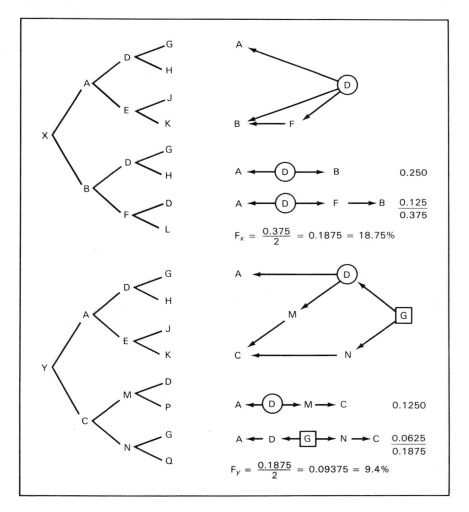

Figure 14.8 Inbreeding coefficients of collateral relatives X and Y.

25 percent greater because of the inbreeding in the common ancestor. The actual calculated relationship is estimated to be 0.25(1.25) = 0.3125, or 31.25 percent. Due to chance, the sire S has almost one-third of the same genes as does the dam D. The fact that the ancestor B was inbred, and thus more homozygous than average, increased the probability that more of the same genes would be passed on to his descendents.

Inbreeding in one or both of two related individuals tends to reduce the relationship between them. Inbreeding causes them to be more homozygous, which reduces the variability of their genotypes. This in turn decreases the chances of their having the same genes by about the average of their two inbreeding coefficients. In the example in Figure 14.8, both X and Y

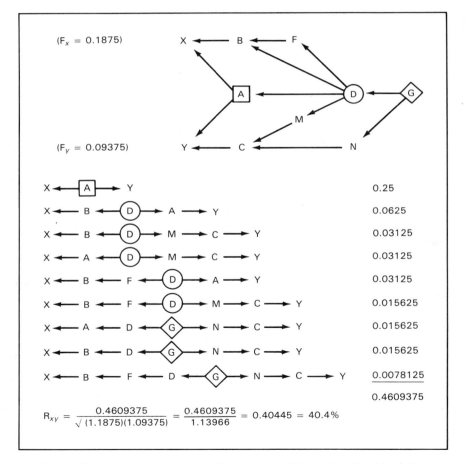

Figure 14.9 Arrow diagram, path coefficients, and coefficient of relationship of collateral relatives X and Y.

are inbred. The new formula for the relationship coefficient is such cases is

$$R_{xy} = \frac{\Sigma(\frac{1}{2})^n (1 + F_a)}{\sqrt{(1 + F_x)(1 + F_y)}}$$

where R_{xy} is the coefficient of relationship between X and Y, and F_x and F_y are the inbreeding coefficients of X and Y, respectively. Figure 14.9 includes the arrow diagram of the combined pedigrees in Figure 14.8, and the list of pathways and path coefficients used to calculate relationship and inbreeding coefficients. When one or both of two relatives are inbred, their relationship is reduced by about the average of their relationship coefficients. In the formula above, the averaging procedure is accomplished by obtaining the square root of the product of 1 plus the inbreeding coefficients of X and Y, respectively. In effect, the relationship coefficient will be reduced by a factor of approximately 14 percent, from about 46 percent for the unadjusted relationship, to about 40 percent.

STUDY QUESTIONS AND EXERCISES

14.1. Define each term.

ancestor   linebreeding
collateral relative   pedigree
common ancestor   prepotency
descendent   relationship
inbreeding   relationship coefficient
inbreeding coefficient

14.2. What is the coefficient of relationship between a woman and her **(a)** half sister? **(b)** aunt? **(c)** first cousin? **(d)** mother? **(e)** husband? **(f)** grandfather? **(g)** niece?

14.3. What is the genetic effect of inbreeding? Explain.

14.4. How can inbreeding change genotypic frequencies but not change allelic frequencies?

14.5. What is the most extreme form of inbreeding possible in animals?

14.6. How is linebreeding similar to inbreeding in general?

14.7. How is linebreeding different from inbreeding in general?

14.8. What is the usual phenotypic effect of inbreeding?

14.9. How are recessive genes "uncovered" with inbreeding?

14.10. What kinds of traits are affected most by inbreeding?

14.11. What kinds of traits are affected least by inbreeding?

14.12. How can inbreeding be used to increase prepotency?

14.13. Why does linebreeding usually have a better reputation among breeders than does inbreeding in general?

14.14. Why is it more practical to use some form of inbreeding among purebred animals than in commercial herds?

14.15. If a male were bred to a number of his daughters and produced excellent offspring, what possible explanation could be given for these results?

PROBLEMS

14.1. Assign one set of numbers to the individuals in the following pedigrees. Begin with number 1 for Monitor, 2 for Caroline, 3 for Flamboyant, and so on.

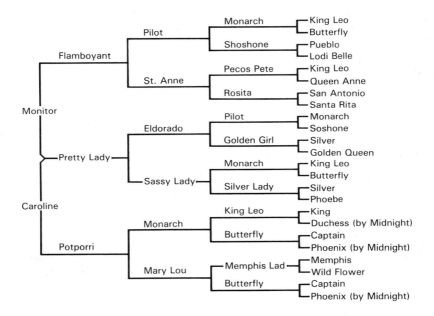

14.2. Calculate the inbreeding coefficient of Monarch.

14.3. Calculate the inbreeding coefficient of Monitor.

14.4. Calculate the inbreeding coefficient of Caroline.

14.5. Calculate the relationship coefficient between Monitor and Monarch (note that both are inbred).

14.6. Calculate the relationship coefficient between Caroline and Monarch (note that both are inbred).

14.7. Calculate the inbreeding coefficient of Pretty Lady.

14.8. Calculate the relationship coefficient between Monitor and Caroline (note that two common ancestors and both Monitor and Caroline are inbred).

REFERENCES

BERESKIN, B., C. E. SHELBY, K. E. ROWE, W. E. URBAN, JR., C. T. BLUNN, A. B. CHAPMAN, V. A. GARWOOD, L. N. HAZEL, J. F. LASLEY, W. T. MAGEE, J. W. MCCARTHY, and J. A. WHATLEY, JR. 1968. Inbreeding and swine productivity traits. *Journal of Animal Science* 27 : 339.

BURGESS, J. B., N. L. LANDBLOM, and H. H. STONAKER. 1954. Weaning weights of Hereford calves as affected by inbreeding, sex and age. *Journal of Animal Science* 13 : 843.

DINKEL, C. H., D. A. BUSCH, J. A. MINYARD, and W. R. TREVELLYAN. 1968. Effect of inbreeding on growth and conformation of beef cattle. *Journal of Animal Science* 27 : 313.

FALCONER, D. S. 1981. *Introduction to Quantitative Genetics,* 2nd ed. Longman, Inc., New York.

LASLEY, J. F. 1987. *Genetics of Livestock Improvement,* 4th ed. Prentice-Hall, Inc., Englewood Cliffs, N.J. Chapters 10–12.

SQUIRES, C. E., G. E. Dickerson, and D. T. MAYER. 1952. Influence of inbreeding, age and growth rate of sows on sexual maturity, rate of ovulation, fertilization and embryonic survival. *Missouri Agricultural Experiment Station Bulletin* 494.

WARWICK, E. J., and J. E. LEGATES. 1979. *Breeding and Improvement of Farm Animals,* 7th ed. McGraw-Hill Book Company, New York. Chapter 7.

FIFTEEN

Outbreeding and Heterosis

15.1 INTRODUCTION TO OUTBREEDING

Outbreeding in general is the mating of animals less closely related than the average of the population. In this sense it is the opposite of *inbreeding,* which is the mating of animals more closely related than the average of the population. According to these definitions, any mating between two animals is either inbreeding or outbreeding; the two animals are either related by more or less than the average of the population. In practice, it is not easy to determine average relationship in a population. One rule of thumb is that if two animals do not share a common ancestor for at least five generations, a cross between them would constitute some form or outbreeding. Being related by less than the average of a population is loosely referred to in this discussion as being "unrelated."

At least three types or categories of outbreeding can be described: linecrossing, breedcrossing, and species crossing. *Linecrossing* is the mating of unrelated animals within the same breed, or from two different lines within a breed. Linecrossing is also called *outcrossing. Breedcrossing* is the mating of animals from two different breeds. It is most often called *crossbreeding* and is the most common form of outbreeding practiced. *Species crossing* is the least common form of outbreeding, mainly because animals from different species do not often interbreed. The classic example of a species cross is the mule, a cross between a stallion and a jenny.

15.2 EFFECTS OF OUTBREEDING

15.2.1 Genetic Effect

The genetic effect of outbreeding is to increase the proportion of genes that are heterozygous and reduce the proportion that are homozygous. This effect is the opposite of that of inbreeding. As the proportion of heterozygous pairs of genes increases with outbreeding, homozygosity increases at a rate such that the frequencies of alleles do not change. The frequencies of genotypes, however, do change.

The degree to which heterozygosity is changed by outbreeding depends in part on the degree to which the animals are related. Crossbreeding is a more extreme form of outbreeding than is linecrossing because two individuals from two different breeds will likely be less related to each other than will two members of the same breed, even though they may be from different lines or families. For this reason, crossbreeding generally results in a more rapid increase in heterozygosity than does linecrossing. Where crosses are possible between animals from different species, the rate of increase in heterozygosity is greater than with crossbreeding.

The term *hybrid* is used quite often in discussions of outbreeding. It was defined in an earlier chapter as an individual that was heterozygous for one or more pairs of genes (i.e., monohybrid, dihybrid, etc.). It can also be used to describe the progeny of the first cross between two species, breeds, or pure lines within a breed. In this context, the "black baldie" calves produced from crossing Herefords with Angus cattle are hybrids. The mule is a species hybrid.

15.2.2 Phenotypic Effect

When unrelated animals are mated together, the progeny tend to perform better than their parents for certain traits, normally those with low heritabilities. The advantage observed among outbred animals is referred to by the general term *hybrid vigor*. It has the opposite effect of inbreeding depression. The actual amount of hybrid vigor can be measured. The quantitative measurement of hybrid vigor, called *heterosis,* is defined as the percent increase in performance of outbred animals over the average of their parents.

To illustrate the definition of heterosis, an example will be used of weaning weight in sheep. In a certain experiment, purebred Hampshire lambs weighed an average of 58.3 lb at weaning, while purebred Suffolks weighed an average of 73.3 lb. The average of the two purebreds was 65.8 lb. When the two breeds were crossed, the Hampshire–Suffolk hybrid lambs weighed an average of 72.8 lb at weaning. The percent heterosis is calculated according to the following formula:

$$\text{heterosis} = \frac{[72.5 \text{ lb (average of hybrids)} - 65.8 \text{ lb (average of purebreds)}] \times 100}{65.8 \text{ lb (average of purebreds)}}$$

$$= \frac{6.7 \text{ lb} \times 100}{65.8 \text{ lb}} = 10.2\%$$

Notice that the comparison is made between the average values of the purebred parents and the average of hybrid progeny. Heterosis exists if the average of the hybrids significantly exceeds the average value of the purebred parents, not if it exceeds the value of the superior parent breed. In this example, the average performance of the hybrids is below that of the better parent breed.

15.3 FACTORS AFFECTING HETEROSIS

The amount of heterosis obtained from outbreeding is dependent on a number of factors, including the genetic diversity of the parents and the type of gene action involved. As will be seen, some factors are related to others.

15.3.1 Heterosis and Genetic Diversity of Parents

The rate of increase in heterozygosity obtained from outbreeding depends in part upon how genetically dissimilar the parents are. The more unrelated two animals are, the fewer genes they will have in common and the greater the heterozygosity in their outbred progeny.

Suppose that a bull from one Hereford line were bred to a cow from a relatively unrelated Hereford line. A small increase in heterozygosity in the linecross calf would be expected over the average heterozygosity of the two parents. The genetic diversity of members of the two lines would be greater than that of the average of the Hereford breed in general. If Hereford bulls were mated to Angus cows, and Angus bulls to Hereford cows, a greater increase in heterozygosity would occur in the F_1 hybrid calves than occurred in the linecross Herefords (or linecrossed Angus, for that matter). A greater percentage of heterosis would be expected in the performance of the crossbred calves compared to the average of linecross calves from the different breeds.

The common origins of British and other European breeds of cattle and the humped cattle of India are more distant that those of the various

TABLE 15.1   Effect of Genetic Diversity on Calving Rate in Beef Cattle

Breeds in Cross (Sire Breed Listed First)	Calving Percentage		Advantage of Cross	Percent Heterosis
	Cross	Pure		
Angus × Hereford	68.2	65.4	2.8	4.3
Hereford × Angus	71.4	65.4	6.0	9.2
Means	69.8	65.4	4.4	6.7
Hereford × Brangus	67.1	65.6	1.5	2.3
Brangus × Hereford	77.1	65.6	11.5	17.5
Means	72.1	65.6	6.5	9.9
Angus × Brahman	79.7	66.5	13.2	19.8
Brahman × Angus	77.4	66.5	10.9	16.4
Hereford × Brahman	86.4	66.8	19.6	29.3
Brahman × Hereford	84.7	66.8	17.9	26.8
Means	82.1	66.7	15.4	23.1

Source: Turner et al. (1968).

European breeds themselves. For this reason, crosses between Brahman cattle and either Herefords or Angus generally result in greater amounts of hybrid vigor than do crosses between Angus and Herefords. The data in Table 15.1 illustrate this very well. The average heterosis in calving percentage involving Angus and Hereford cattle is 6.7 percent, while that involving crosses of Brahmans with Herefords and Angus is 23.1 percent.

Since Brangus cattle were developed from Brahman and Angus foundation stock, the genetic diversity between Hereford cattle and cattle of the Brangus breed would be expected to be intermediate to that of Hereford and Angus, and Hereford and Brahman. The amount of heterosis resulting from crosses between Hereford and Brangus should therefore be intermediate to that of Hereford–Angus and Hereford–Brahman crosses. According to the data in Table 15.1 showing a heterosis of 9.9 percent for crosses between Brangus and Herefords, this is true.

15.3.2 Effects of Gene Action and Heritability

The kinds of gene action having the greatest effect on heterosis are the kinds classified as nonadditive. These include dominance and recessiveness, overdominance and epistasis. The examples in Figure 15.1 illustrate this. Only a few gene loci are used in the illustrations, but the effect would be similar regardless of the number of loci involved. The actual effects of heterosis are probably due to a combination of several kinds of nonadditive gene action. As suggested in the illustration, additive genes have very little or no effect on heterosis.

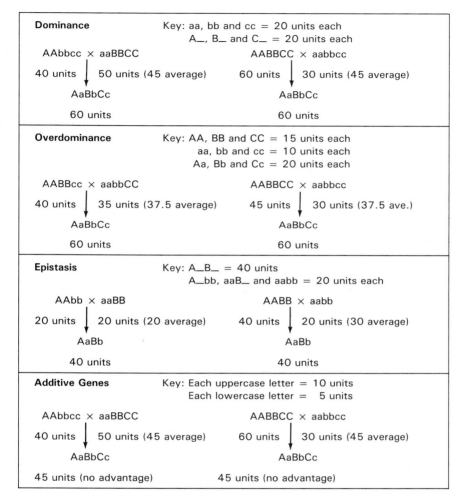

Figure 15.1.   Hypothetical examples showing effects of several kinds of gene action on heterosis.

Traits with higher heritabilities (in the narrow sense) tend to be controlled more by additive genes than by nonadditive genes. Since heterosis tends to be affected less by additive genes, it follows that, generally, the higher the heritabilities among traits, the lower the percent heterosis (see Table 15.2). Traits with low heritabilities, such as litter size in swine, calving percent, and lambing percent, generally exhibit quite a bit of heterosis. Litter size in swine, for example, may express from 8 to 20 percent heterosis,

TABLE 15.2   General Effects of Heritability on Heterosis

Heritability of Trait	Effect of Additive Genes	Effect of Nonadditive Genes	Amount of Heterosis from Outbreeding
Low	Small	Large	Large
Medium	Intermediate	Intermediate	Medium
High	Large	Small	Small

depending on the number and combination of breeds used in the program. Carcass traits, such as area of the loin eye, are highly heritable and express almost no heterosis from crossbreeding.

15.4 EFFECTS OF OUTBREEDING IN VARIOUS FAMILY MEMBERS

Numerous experiments have been conducted in swine, cattle, and sheep to determine the effects of outbreeding in both purebred and commercial herds. The impact and relative importance of heterosis in progeny, dams, and sires are examined in the next few paragraphs.

15.4.1 Heterosis in Progeny

Heterozygosity results in an increase in survival rates of embryos and fetuses in essentially all species, which in turn results in more live births. Litter size at birth in swine when only the pigs are crossbred may be increased by one-half to one pig per litter on the average. This increase is due to heterosis only in the pigs since the parents are not crossbred. For example, suppose that a pure line of Hampshire swine averaged 7.6 pigs born per litter, and a pure line of Durocs averaged 7.4 pigs per litter. Crosses between Hampshires and Durocs would be expected to produce litters that averaged over 8 pigs in number.

Similar advantages can be expected in birth rates in cattle and sheep. Since they are not litter-bearing species, birthrates in cattle and sheep are measured in terms of the number of calves or lambs per 100 females bred. Increases in lambing and calving rates of 5 to 10 percent are common when crossbred progeny are produced from straightbred females.

There is also an advantage in survival rates to weaning, growth rates, and efficiency of gains of outbred progeny compared to their straightbred parents. The term *straightbred* refers to animals that are essentially purebred but are not registered. It does not imply that animals are of lesser quality than purebreds. Table 15.3 includes some typical examples of levels of performance for several traits in swine, along with the percent heterosis for each trait.

Column 3 shows the performance levels due to crossbreeding only in the pigs with the actual advantage over the purebreds listed in parentheses. A two-breed cross is designed to produce crossbred pigs from purebred or straightbred dams. The percent heterosis expected in the various traits from a two-breed cross are listed in column 4. Litter weights at weaning due to heterosis in the progeny show an advantage of 107 lb over the purebreds, or 31 percent. These weights are the product of the number of pigs per litter and the pig weights at weaning (5.6 pigs × 61 lb for purebreds versus 6.7 pigs × 67 lb for crossbreds). Litter size at weaning shows 20 percent heterosis and pig weight shows an advantage of 10 percent. Litter weight at weaning shows the combined effects of litter size and pig weight.

15.4.2 Heterosis in Dams

Normal or usual ovulation rates vary widely among mammals. Swine and other litter-bearing species may ovulate from 10 to 15 ova during one estrous cycle. Mares generally have the lowest ovulation rates among domestic animals, averaging less than one per cycle. Whether ovulation rates are measured in terms of number per female as in litter-bearing animals, or in number per 100 females as in mares and cattle, crossbred females ovulate at a greater rate.

The number of progeny born per dam or 100 dams is in part a function of the ovulation rate. It is also affected by the uterine environment and

TABLE 15.3 Typical Examples of Performance and Estimated Percent Heterosis for Several Selected Traits in Swine

(1)	(2)	(3)	(4)	(5)	(6)
	Purebred Pigs from Purebred	Crossbred Pigs from Purebred	Percent Heterosis	Crossbred Pigs from Crossbred	Percent Heterosis
Trait	Dams	Dams	(Two-Breed)	Dams	(Three-Breed)
Number of pigs born	7.5	8.1 (0.6)[a]	8	9.0 (1.5)[a]	20
Number of pigs weaned	5.6	6.7 (1.1)	20	8.4 (2.8)	50
Pig weight at weaning (lb)	61	67 (6)	10	68 (7)	11
Litter weight at weaning (lb)	342	449 (107)	31	571 (229)	67

[a]Advantage over the purebreds.

the inherent ability of the embryos and fetuses to survive. It was indicated earlier that crossbred fetuses have a higher survival rate than do purebred fetuses. Crossbred dams produce more progeny at birth, in part due to an improvement in the uterine environment compared to purebred females.

Number of pigs born from crossbred sows (Table 15.3, column 5) is usually greater than the number born from purebred sows whether pigs from the latter are crossbred or not. The advantage of 1.5 pigs shown in the table is due to a combination of the heterosis in both sows and pigs. About 0.6 of that advantage was the result of heterosis only in the pigs. The difference of 0.9 pig can be attributed to heterosis in the sow.

During the period from birth to weaning, the crossbred dam provides a further advantage over purebreds by producing more milk. The amount of milk produced by a female is probably the greatest single factor affecting weaning weights and litter weights at weaning. Notice in Table 15.3 that in addition to the 107-lb advantage in the weight of the litter due to heterosis in the pigs, when sows are also crossbred the advantage is about 229 lb. The difference between these numbers, 122 lb, is the advantage due to heterosis in the sow.

Outbreeding in cows and ewes has effects similar to those in swine. For example, in one experiment designed to study the heterosis in weaning weights in beef cattle, the average calf weights for purebred Herefords, Brahmans, and Brangus were 346, 392, and 440 lb, respectively. The three-breed average was 386 lb. The overall average weaning weight for all possible two-breed crosses among the three breeds was 425 lb. Again, with two-breed crosses, the calves were crossbred, the dams were not. Heterosis in this experiment was 10.1 percent [(425 − 386)100/386 = 10.1 percent].

The average weaning weights of calves from all possible three-breed crosses was 460 lb. In these situations, two-breed cross dams were bred to a third breed of bull. The calves would contain one-half their genetic makeup from one breed and one-fourth each from the two other breeds. The average heterosis due to crossbreeding in the cows and calves was 19.2 percent [(460 − 386)100/386 = 19.2 percent]. The two-breed calves weighed an average of 39 lb more than purebred calves; the three-breed calves were 74 lb heavier. The difference of 35 lb can be attributed to heterosis in the dams.

The preceding examples in swine and beef cattle illustrate the advantages of crossbreeding in the commercial production of pork and beef. Heterosis can be demonstrated due to crossbreeding in dams as well as to crossbreeding in their progeny.

15.4.3 Heterosis in Sires

The contribution made by males of domestic species to their progeny is generally limited to one-half the genetic makeup carried in their spermatozoa. Not only do females contribute one-half the genes to their progeny, they also provide the uterine environment during pregnancy, care and protection prior to weaning, and the majority of the food supply before weaning. As we have seen, crossbreeding in the dams enhances each of these added contributions.

Since domestic animal sires contribute only one-half the genotype of their progeny, there is usually little or no advantage to using crossbred sires in a breeding program. From a selection standpoint, there may even be some disadvantage. Any advantage in performance expressed due to heterosis will not be transmitted in the gametes. This is based on the principle described earlier that heterozygotes do not breed true. Any advantage obtained from the interaction between alleles in the heterozygous state must be reestablished each generation by outbreeding.

Because of the inability to transmit hybrid vigor, superior performance expressed by crossbred sires must be discounted by about the percent heterosis of the respective traits to obtain a measure of their true breeding value. Because of hybrid vigor, crossbred sires are not as prepotent as inbred sires, so they tend to "look better than they breed." Superiority in performance in inbred males is more likely to be fixed in the homozygous state and not the result of heterozygosity. Inbred males can therefore be expected to "breed better than they look." Considering the precautions presented concerning the use of outbred sires, their use should be evaluated on an individual basis.

15.5 COMMERCIAL CROSSBREEDING PROGRAMS

It has been shown that the greatest benefits from crossbreeding are obtained in dams and progeny for traits of low to medium heritabilities. One function of the commercial breeder is to maximize the advantages of heterosis in the efficient and economical production of food and fiber from livestock.

15.5.1 Terminal Crosses

Minimal advantages can be obtained from terminal two-breed crosses in commercial herds of swine and beef cattle. This system is most useful when high-quality straightbred females are bred to superior males. All F_1 progeny are sold, usually for feedlot purposes. Replacement females must either be purchased or produced in separate breeding herds. It is a better system than straightbreeding, because it takes advantage of heterosis in the progeny. Since it does not take advantage of heterosis in dams, it is generally not as good as a three-breed crossbreeding program.

Maximum effects from heterosis can be obtained when 50 percent of the genes in both dams and progeny come from a single breed source and the other 50 percent come from a second breed or combination of breeds. A three-breed terminal crossbreeding program takes advantage of this situation (Figure 15.2). The best-quality F_1 females available are bred to a male of a third breed. Maximum benefits from heterosis in the dams and in the progeny are possible. Genetic diversity of the breeds is also a factor; the more diverse the genetic relationship among the breeds, the greater the heterosis.

As was the case with the two-breed terminal system, all progeny from a three-breed terminal crossbreeding program are sold as feeder or stocker animals. High-quality crossbred females must be purchased to replace the dams that are culled. Herein lies one possible disadvantage of terminal breeding systems; breeders cannot take advantage of selection of females from within the herd. On the other hand, in some herds, especially very small beef cattle herds, greater advantage can sometimes be obtained from a three-breed terminal crossbreeding program than that obtained from a three-breed rotational system.

15.5.2 Quality versus Maximum Heterosis

One purpose of a commercial crossbreeding program is to provide maximum, or optimum, benefits from heterosis generation after generation. As already pointed out, maximum heterosis can be obtained when

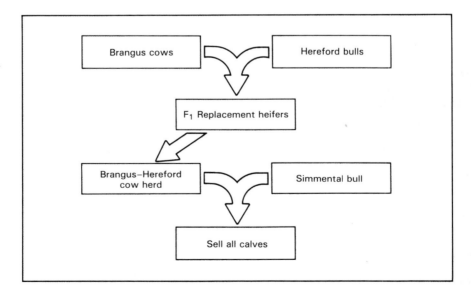

Figure 15.2   Example of a three-breed terminal crossbreeding program.

two-breed F_1 females are bred to a male of a third breed. To obtain progeny where exactly 50 percent of their genetic makeup originates from one breed source, the three-breed females must be bred to a fourth breed. This is certainly possible and is to be recommended if males of a fourth breed can be obtained that possess quality in traits equal to that of the three breeds already in use.

The following illustration is presented to demonstrate how deviations from the 50 : 50 ratio of breed composition in crossbred animals can affect heterosis. The 50 : 50 ratio means that 50 percent of the genes come from one breed source while the other 50 percent come from one or more other sources. Suppose that weaning weights of crossbred calves produced from breeds A and B are 6 percent heavier than average purebred calves from breeds A and B. When A/B cows are bred to bulls of breed C, assume that another 6 percent advantage in weaning weights is obtained due to heterosis in the cows. The total advantage in three-breed cross calves would be about 12 percent.

If the A/B cows were backcrossed to one of the parent breeds, say to breed A, three-fourths of the genes in the calves would be from breed A and one-fourth would be from breed B. The A/B cows would contribute their 6 percent advantage in heterosis due to their 50 : 50 gene source. However, since three-fourths of the genetic composition of the calves is from a single breed, they would not be as heterozygous as calves with exactly one-half their genes from one source. According to the data presented in Table 15.4, the reduction in heterosis in animals of a 75 : 25 ratio of breed sources is about 50 percent of the maximum obtained when the sources are 50 : 50.

TABLE 15.4   Effect of Proportion of Genes from One Breed on Maximum Heterosis Possible from Crossbreeding

Proportion of Genes from a Single Breed		Proportion of Maximum Heterosis Possible[a]	
Percent	Fraction	Fraction	Percent
100	All	None	0
93.75	$\frac{15}{16}$	$\frac{1}{8}$	12.5
87.5	$\frac{7}{8}$	$\frac{1}{4}$	25
81.25	$\frac{13}{16}$	$\frac{3}{8}$	37.5
75	$\frac{3}{4}$	$\frac{1}{2}$	50
68.75	$\frac{11}{16}$	$\frac{5}{8}$	62.5
62.5	$\frac{7}{8}$	$\frac{3}{4}$	75
56.25	$\frac{9}{16}$	$\frac{7}{8}$	87.5
50	$\frac{1}{2}$	All	100

[a]According to the formula: $2(100 -$ percent of genes from predominant breed).

In other words, rather than the three-quarter calves contributing 6 percent heterosis, they would be expected to contribute only about 3 percent. The total advantage from heterosis in the cows and calves would be about 9 percent instead of 12 percent as cited for the three-breed cross calves.

It must be remembered that benefits of heterosis come from the increase in the performance of crossbred animals over that of the average performance of purebreds used in the crosses. If poor-quality straightbreds or purebreds are used in a crossbreeding program, the performance of the crossbreds could be inferior to the performance of more superior purebreds. The next example is presented to illustrate how the level of performance of pure breeds can affect the relative advantages of crossbreeding.

Assume that the weaning-weight averages for the three breeds A, B, and C in the previous example are 450, 440, and 405 lb, respectively. The average of breeds A and B is 445 lb. With 6 percent heterosis, the weaning weights of the A/B cross calves would be expected to average about 472 lb. Crossing the A/B cows to bulls of breed C would be expected to produce calves with an average of 476-lb weaning weights, a 12 percent advantage over the 425-lb average of the three breeds (25 percent A + 25 percent B + 50 percent C). Because the average of breed A is so much higher than that of breed C, if the A/B cows were bred back to a bull of breed A, the expected average weaning weight of the 75 : 25 calves would be about 488 lb. A 9 percent advantage over the average of the two best breeds is better than a 12 percent advantage over the average of three breeds when the third breed is mediocre.

The preceding two examples were presented to illustrate that the ultimate advantage of heterosis depends not only on the number, percentage, and diversity of the breed sources, but upon the inherent quality of the purebred or straightbred foundation stock.

15.5.3 Rotational Crossbreeding Programs

These breeding programs make it possible to take advantage of two very important tools of livestock improvement: crossbreeding and selection. Crossbreeding makes it possible to take advantage of available heterosis. The rotational part of the system provides two things: a way to maintain heterozygosity in dams and progeny generation after generation, and a way to take advantage of selection of female stock from within the herd. The most common rotational systems make use of two or three breeds, but four or more breeds could be used.

The two-breed rotational crossbreeding program is sometimes called a *two-breed crisscross program* (see Figure 15.3). It is useful when male breeding stock of desired quality and characteristics is available in only two breeds. This system can be illustrated using beef cattle of the Angus and

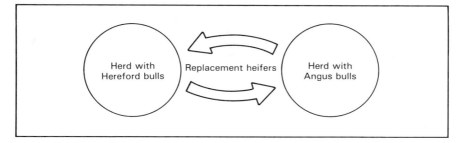

Figure 15.3   Example of the two-breed crisscross (rotation).

Hereford breeds. F_1 heifers from the crossing of these two breeds would be bred back to either Angus or Hereford bulls. One-fourth of the genes in these second-cross progeny would be from one breed with three-fourths from the other breed. The effect of heterosis due to their own heterozygosity would be less than that of the 50 : 50 cross calves; however, they would benefit from having crossbred dams, whereas the 50 : 50 cross calves would not.

Heifers that possessed 75 percent Hereford breeding and 25 percent Angus breeding, for example, would be bred back to Angus bulls, resulting in progeny with 37.5 percent Hereford background and 63.5 percent Angus background. After several generations of breeding in such a program, the proportion of genes from the two breeds would approach a ratio of two-thirds from one breed and one-third from the other. The proportion of the maximum heterosis obtained in a two-breed crisscrossing system would approximate two-thirds.

The three-breed rotational crossbreeding program has come to be somewhat of a standard to which other crossbreeding programs are compared, especially in commercial swine and beef cattle herds. The first two generations of this system are essentially identical to those of the three-breed terminal crossing system. The first cross generation are F_1 hybrids consisting of genes from two pure breeds. The second generation is produced by breeding the F_1 hybrids to a third breed of sire. The example in Table 15.5 shows the variation in percentage of genes from each of three breeds of swine for several generations. As described earlier, the second generation is where maximum heterosis is obtained in both dams and progeny.

The third generation is produced by mating gilts with 50% Yorkshire, 25% Duroc, and 25% Hampshire back to a Duroc boar. The pigs from this cross will have obtained about 62.5 percent of their genes from the Duroc breed. Because the percentage of genes from one breed is more than 50 percent, the amount of heterosis obtained in this generation will be less than the previous one. Refer again to Table 15.5. For the next several generations

TABLE 15.5   Three-Breed Rotational Crossbreeding System in Swine

Generation Number	Percent of Genes from Breed of Dam			Breed of Boar to Which Sow Will Be Bred
	Yorkshire	Duroc	Hampshire	
0	100			Duroc
1	50	50		Hampshire
2	25	25	50	Yorkshire
3	62.5	12.5	25	Duroc
4	31.25	56.25	12.5	Hampshire
5	15.625	28.125	56.25	Yorkshire
6	57.8	14.1	28.1	Duroc
7	28.9	57.0	14.1	Hampshire

of rotational crossing, the percentage of genes from one breed source will vary between 55 and 60 percent. The proportion of genes from the three breeds will eventually approach a ratio of 4 : 2 : 1, or approximately 57.1, 28.6, and 14.3 percent, respectively. The percent of maximum heterosis can be estimated at about 86 percent using the formula given in Table 15.4.

The percent of maximum heterosis obtained in a four-breed rotation will ultimately approach about 93 percent. However, most experiments have shown no particular advantage from the use of four breeds compared to three. The key to whether more than three breeds would be used in the rotation would depend on the availability of sires with the desired characteristics and quality.

The optimum benefits of using a three-breed rotational crossbreeding system are realized when the herd is large enough to justify the use of at least three sires. The larger herd can be divided into thirds, depending on the breed of sire to which the females are to be bred. Replacement females from one generation can be rotated to the herd with the breed of sire to which they will be bred where they will remain until they are culled.

Using the three breeds of swine in Table 15.5 as an example, there would be three breeding herds, each with one breed of sire. To illustrate, replacement gilts from generations two and five, having been sired by a Yorkshire boar, would be moved to the breeding herd having only Duroc boars (see Figure 15.4). The replacement females from this herd would be moved next to the "Hampshire" herd, and so on. In this way the optimum percentage of genes from the three respective breeds can be maintained in the entire herd.

If the overall herd was large enough to justify only one sire, as in the case of a 30-cow commercial beef cow operation, maximum benefits from heterosis in a three-breed rotation may not be possible in each generation. The herd sire of a one-bull herd should probably be replaced every two years to avoid breeding a bull to his daughters. Within the herd there would

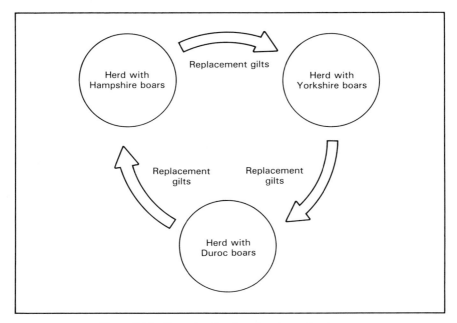

Figure 15.4   Example of a three-breed rotational cross.

be cows with various percentages of genes from the three breeds. The breed of choice for a replacement sire would be the one represented by the least percentage in the greatest number of cows. Some cows might contain $\frac{5}{8}$ or $\frac{11}{16}$ of their genes from one breed, but because they were few in number, they might have to be bred to a bull of that breed. The percent heterosis in their calves would therefore be reduced. Because of this situation, a three-breed terminal crossing program could be of greater benefit in small herds. At least it would be possible to take advantage of the maximum benefits of heterosis. Since F_1 hybrid replacement females would have to be purchased, the quality of those females would be the determining factor in deciding whether the three-breed terminal or the three-breed rotational system would be more beneficial.

If the management program is such that a breeder can make use of AI sires, it is possible to maintain a crossbreeding program involving three breeds and obtain the maximum benefits of both heterosis and selection.

15.5.4 Combination Rotational–Terminal Cross Programs

This program has important applications in large commercial herds, particularly with beef cattle (see Figure 15.5). Up to about 50 percent of the females are used in a two-breed crossing program to produce a source of

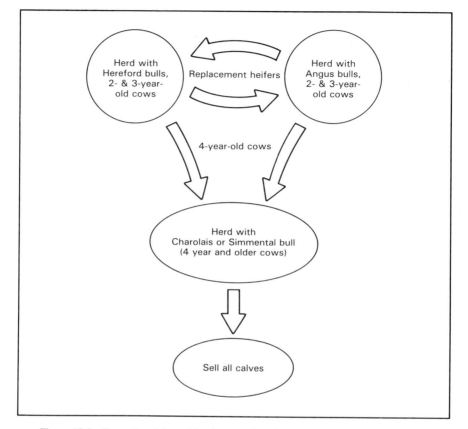

Figure 15.5 Example of a combination rotational–terminal crossbreeding program.

crossbred replacements. This part of the herd would involve the smaller and younger cows. Older cows, and larger cows and heifers, are bred to terminal cross sires of the larger breeds to obtain maximum benefits of heterosis.

As an example, Hereford- and Angus-bred cows could be used in a two-breed rotational system to produce crossbred replacements. The replacements would be bred back to Hereford or Angus bulls through three or four calf crops. The older Hereford–Angus cows would then be bred to Charolais or Simmental bulls.

This program has the advantage over a terminal crossing system in that replacement heifers can be selected from within the herd. However, it requires that nearly all heifers be kept for breeding. This makes it even more important to maintain a near-100 percent calving rate. Only about 50 percent of maximum heterosis can be obtained in the rotational herd, but 100 percent is obtained in the terminal-sire herd. The fact that maximum heterosis is not obtained in the rotational breeding herd is compensated for, at least in part, by the use of the larger growth sires in the terminal herd.

15.5.5 Conclusions Concerning Crossbreeding

The advantages of crossbreeding are generally associated with traits of low to medium heritability. In swine, litter size and litter weight at weaning are improved considerably. This advantage comes from increased vigor in the pigs and from vigor and increased milk production in sows. In some cases, slightly faster rates of gain and increased efficiency are observed during the period from weaning to market weight.

Most of the direct advantage from crossbreeding in beef cattle is realized by the time calves have been weaned. More calves are born per 100 cows bred when the calves are crossbred, as well as when cows are crossbred compared to purebred and straightbred calves and cows. Calves weigh more at weaning when they are crossbred. Calves weaned by crossbred cows also weigh more, due mainly to increased milk production of the cows. There is also evidence to indicate that crossbred cows may be better mothers.

Some of the same advantages of crossbreeding observed in beef cattle and swine are also noted in sheep. Advantages from heterosis are apparent in reproductive efficiency, weaning weights, milk production, mothering ability, and wool traits.

From a practical standpoint, crossbreeding has not been particularly advantageous as a tool for improving dairy cattle, especially as a method for increasing overall production of milk or butterfat. Milk production does express heterosis, but even the highest-producing crossbreds will generally not outproduce the best Holsteins. To illustrate, suppose that a line of Holsteins produce at an average rate of 24,000 lb of milk per year, and a line of Ayrshires produce at the rate of 20,000 lb. The average of the two breeds is 22,000 lb. Even if heterosis were as high as 6 percent, the crossbreds would only be expected to produce about 23,320 lb of milk on the average, which is below the average for the Holsteins in this example.

For other traits in dairy cattle, such as calving rate, preweaning growth rate, and increased vigor in cows and calves, crossbreeding can provide advantages. Some of the benefits of heterosis can be realized in purebred or straightbed herds from linecrossing.

STUDY QUESTIONS AND EXERCISES

15.1. Define each term.

breedcrossing	hybrid	outcrossing
crossbreeding	hybrid vigor	prepotency
genetic diversity	linecrossing	purebred
heterosis	outbreeding	straightbred

15.2. What is the genetic effect of outbreeding?

15.3. What is the difference between the effect of linecrossing and crossbreeding on genotype?

15.4. What is the effect of outbreeding on allelic frequencies?

15.5. What form of outbreeding results in the greatest amount of heterosis?

15.6. What is the relationship between sizes of heritability estimates among traits and the amount of heterosis possible?

15.7. List some biological factors that contribute to the number of pigs produced at weaning.

15.8. How does outbreeding affect the number of pigs at weaning age?

15.9. What is probably the single most important reason why crossbred mothers produce heavier progeny at weaning?

15.10. What disadvantage may be involved with the use of crossbred sires?

15.11. What is meant by a "terminal" crossbreeding program?

15.12. What is meant by a "rotational" crossbreeding program?

15.13. Describe one possible way to operate a three-breed rotational crossbreeding system in a three-sire herd.

15.14. What advantage does a rotational crossbreeding system have over a terminal system?

15.15. How might a terminal three-breed crossbreeding system in a one-bull herd possibly result in greater total benefits than a rotational system?

15.16. Why is crossbreeding not commonly practiced in commercial dairy herds?

PROBLEMS

15.1. Assume that the annual average milk production of one pure line of dairy cattle is 17,000 lb, and that of another line is 18,000 lb. Calculate the percent heterosis if the outbred daughters produce at an annual rate of 18,400 lb.

15.2. Assume average birthweights for Hereford cattle to be 72 lb, and those of Brahmans to be 86 lb. Calculate the percent heterosis if birthweights produced from reciprocal crosses of Herefords and Brahmans average 82 lb.

15.3. Suppose that the average litter size at birth for selected lines of Berkshire, Poland China, and Spotted Swine are 7.2, 7.4, and 7.8, respectively. Assume heterosis for two breed crosses to be 6 percent and heterosis for three breed crosses to be 17 percent. **(a)** Calculate the predicted average litter size for all crosses between Berkshires and Spotted Swine. **(b)** Calculate the predicted average litter size of all crosses between Spotted Swine and Poland Chinas. **(c)** Calculate the predicted average size of litters produced from all possible three-breed crosses.

15.4. The average calving rate among three pure breeds of beef cattle being used as the basis for a rotational crossbreeding program is 83 percent. The average weaning weight of the three breeds is 440 lb. The average calf weight per cow is calculated as 365 lb (440 lb × 0.83). The average calving rate of all three-breed crosses is 88 percent, and the average weaning weight is 491 lb. Calculate the percent heterosis for **(a)** calving rate, **(b)** weaning weight, and **(c)** average calf weight per cow.

15.5. The genetic makeup of one group of feeder pigs originated from $\frac{1}{2}$ Duroc, $\frac{1}{3}$ Chester White, and $\frac{1}{6}$ Landrace. A second group of pigs consisted of $\frac{1}{8}$ Duroc, $\frac{3}{16}$ Chester White, and $\frac{11}{16}$ Landrace. **(a)** Which group of feeder pigs would be expected to produce the lesser heterosis? **(b)** What percentage of the maximum possible amount of heterosis will the lesser group produce (see Table 15.4)?

REFERENCES

CUNDIFF, L. V., K. E. GREGORY, and R. M. KOCH. 1974. Effects of heterosis on reproduction in Hereford, Angus and Shorthorn cattle. *Journal of Animal Science* 27 : 336.

HOHENBOKEN, W., and P. E. COCHRAN. 1976. Heterosis for ewe lamb productivity. *Journal of Animal Science* 42 : 819.

LASLEY, J. F. 1987. *Genetics of Livestock Improvement,* 4th ed. Prentice-Hall, Inc., Englewood Cliffs, N.J. Chapter 13.

MILLER, K. P., and D. L. DAILEY. 1951. A study of crossbreeding in sheep. *Journal of Animal Science* 10 : 462.

SELLIER, P. 1976. The basis of crossbreeding in pigs: a review. *Livestock Production Science* 3 : 203.

TURNER, J. W., B. R. FARTHING, and G. L. ROBERTSON. 1968. Heterosis in reproductive performance of beef cows. *Journal of Animal Science* 27 : 336.

WARWICK, E. J., and J. E. LEGATES. 1979. *Breeding and Improvement of Farm Animals,* 7th ed. McGraw-Hill Book Company, New York. Chapter 8.

WILTBANK, J. N., K. E. GREGORY, J. A. ROTHLISBERGER, J. E. INGALLS, and C. W. KASSON. 1967. Fertility in beef cows bred to produce straightbred and crossbred calves. *Journal of Animal Science* 26 : 1005.

YOUNG, L. D., R. K. JOHNSON, and I. T. OMTVEDT. 1976. Reproductive performance of swine bred to produce purebred and two-breed cross litters. *Journal of Animal Science* 42 : 1133.

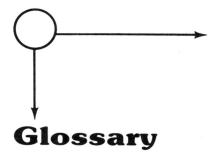

Glossary

Aberration An abnormality, usually associated with malformations in chromosomes.

Acrocentric Refers to a chromosome with a nearly terminal centromere.

Albinism A genetic condition in animals characterized by a complete absence of pigmentation in the skin, hair, and eyes.

Alleles (Allelomorphs) Genes that can occupy the same loci on homologous chromosomes, but have different effects.

Anaphase The stage of mitosis where centromeres of doubled chromosomes divide and separate, producing two sets of chromosomes that migrate to opposite sides of the cell prior to the completion of cell division.

Ancestor An individual from whom an animal or person is descended.

Asexual Without sex.

Autosomes All chromosomes except the sex chromosomes.

Avian Refers to birds.

Backcross A mating of an individual back to a parental type or breed.

Barred A trait in chickens involving alternating white and dark stripes on the feathers.

Bay A color pattern in horses involving a brown or reddish-brown body with black mane, tail, and legs from the hocks and knees down.

Binomial A mathematical expression consisting of two terms or values.

Bitch A female dog.

Boar A male hog.

278

Bovine   Refers to cattle.

Breed   A group of animals with certain common characteristics that distinguish them from others of the species, and which they tend to transmit with reasonable consistency.

Breed true   To transmit a trait genetically with reasonable consistency.

Breedcrossing   *See* Crossbreeding.

Buck   A male sheep, goat, deer, or rabbit.

Bull   A male of the cattle species.

Canine   Refers to dogs.

Caprine   Refers to goats.

Carrier   Refers to an animal that expresses a dominant trait, but also carries a recessive gene.

Centriole   A small cytoplasmic organelle located near the nuclear membrane which divides during mitosis and meiosis and forms the centers toward which the chromosomes move during the process of division.

Centromere   A small structure or part of a chromosome that appears to form an attachment for spindle fibers during cell division.

Character   A distinguishing feature or quality; a synonym for trait.

Chestnut   A reddish or reddish-brown color in horses with mane and tail of the same or lighter color; lighter shades are sometimes called sorrel. A horny or callous growth on the inner sides of the legs of a horse.

Chiasma (pl., chiasmata)   The visible connection or crossover between nonsister chromatids observed during prophase I of meiosis.

Chromatid   One strand of a doubled chromosome observed during mitotic prophase or metaphase, or during prophase I or metaphase I of meiosis.

Chromatin   That part of a cell nucleus that stains intensely; usually refers to chromosomes in an indistinguishable form in a nondividing cell.

Chromosome   One of several relatively long threadlike structures found in a cell nucleus; chromosomes are made of protein and DNA and carry the genes.

Cleavage   A kind of division in a developing embryo whereby cells divide mitotically, but their size is reduced with each division until they are as small as typical body cells.

Codominance   A kind of gene action where two or more alleles exhibit approximately equal expression; lack of complete dominance between alleles.

Codon   A set of three nucleotides in the structure of nucleic acid that is specific for a particular amino acid in protein synthesis.

Collateral Relatives   Animals related not by being ancestors or descendants of each other, but by having one or more common ancestors.

Colt   A young male horse.

Common Ancestor   An ancestor that appears in the pedigrees of two or more individuals.

Complete Dominance   A situation where a gene in the heterozygous state expresses itself to the exclusion of the expression of the allele.

Coupling   Refers to a linkage pattern where two dominant genes are linked and two recessive genes are linked.

Crossbreeding   The mating of animals from different breeds; breedcrossing.

Crossing-over   The exchange of parts between nonsister chromatids of homologous chromosomes that occurs during the synapsis stage of meiosis.

Cryotorchidism   A condition in male mammals wherein one or both testes fail to descend into the scrotum.

Culling   The process of eliminating unwanted or poor-quality animals.

Cytoplasm   That part of the contents of a cell located inside the cell membrane and outside the nuclear membrane.

Dam   The female parent.

Deoxyribonucleic Acid (DNA)   The chemical substance of which genes are made.

Descendent   An animal that has received part of its inheritance from another animal known as its ancestor.

Dihybrid   An individual heterozygous for two pairs of genes (e.g., *AaBb*).

Dihybrid Cross   A cross between two dihybrid individuals (*AaBb* × *AaBb*).

Dioecious   Producing either sperm or ova, but not both: separate sexes.

Diploid   Having two complete sets of homologous chromosomes.

Dizygotic Twins   Twins that develop from two separate fertilized ova.

DNA   *See* Deoxyribonucleic Acid.

Doe   An adult female goat, rabbit, or deer.

Dominance   The situation where one member of a pair of alleles expresses itself totally (complete dominance) or in part (partial dominance) over the other; unqualified, it usually implies complete dominance.

Enzyme   A complex protein that activates, facilitates, or regulates certain chemical reactions; an organic catalyst.

Epistasis   A type of gene action where genes at one locus control or modify the expression of genes at a different locus.

Equine   Refers to horses.

F_1   The first-generation progeny of a P_1 cross; usually refers to an individual heterozygous for one or more pairs of genes.

F_2   Second-generation progeny usually produced by crossing two F_1 individuals.

Fertilization   The union of male and female gametes to form an ovum.

Filial   Refers to child or offspring; the meaning of the F in F_1 and F_2.

Filly   An immature female horse.

Foal   A young horse of either sex before weaning.

Gamete   A reproductive cell (sex cell) that contains one-half the genes and chromosomes of the body cells of the parent: an ovum or sperm cell.

Gametogenesis   A process in plants and animals, male and female, involving the production of gamete; ovigenesis or spermatogenesis.

Gene   The smallest unit of inheritance found as a part of a chromosome.

Gene Frequency   The proportion of one gene existing in a population compared to all alleles that exist.

Generation Interval   The average period of time between the birth of one generation and the birth of the next.

Genetic Correlation   A situation where two or more traits are affected by some of the same genes.

Genetic Diversity   Refers to the dissimilarity of genotypes among relatively unrelated animals.

Genetic Drift   A change in gene frequencies due to random fluctuations caused by chance in mating patterns or sampling errors.

Genotype   The genetic makeup of an animal.

Grading Up   A process of livestock improvement where grade or mediocre females are mated to purebred sires.

Half Sib   A half brother or half sister.

Haploid   Refers to a cell with only one set of chromosomes.

Hemizygous   Refers to a genotype composed of only one gene from any series; usually associated with the genotypes of male mammals or female birds with respect to sex-linked genes.

Heritability   The proportion of the total variation for any trait due to genetic causes, and that can be expected to be passed on to the next generation.

Heritability Estimate (H^2 or h^2)   A quantitative measure of heritability, expressed as a percentage or decimal fraction.

Heterogametic   Refers to the sex that possesses two kinds of sex chromosomes, males in mammals (XY) and females in birds (ZW).

Heterosis   A quantitative measure of the amount or proportion of superiority observed in outbred animals compared to their purebred or straightbred parents; a quantitative measure of hybrid vigor.

Heterozygote   An individual whose genotype with respect to any locus on a pair of homologs consists of two different genes (alleles), such as *Aa;* an individual may be heterozygous at more than one locus; heterozygotes produce two kinds of gametes for each pair of heterozygous genes.

Homogametic   Refers to the sex that possesses only one kind of sex chromosome, females in mammals (XX) and males in birds (ZZ).

Homologous Chromosomes   Chromosomes occurring in pairs, one obtained from each parent, usually the same length, having their centromeres in the same position, and having the same gene loci.

Homozygote   An individual whose genotype with respect to any locus on a pair of homologous chromosomes consists of two identical genes, such as *AA* or *aa;* an individual may be homozygous at more than one locus; homozygotes produce only one kind of gamete and thus "breed true."

Hybrid   A heterozygote; the progeny of a mating between two genetically unrelated parents.

Hybrid Vigor   A qualitative term for heterosis.

Hypostatic   The trait whose expression is suppressed or modified in the case of epistatic gene action.

Identical Twins   Twins that develop from a single fertilized ovum that separates into two embryos shortly after fertilization.

Inbreeding   A type of mating involving animals that are more closely related than the average of the population.

Inbreeding Coefficient   A measure of the increase in homozygosity (or decrease in heterozygosity) in an animal due to inbreeding.

Incomplete Dominance   A condition where the phenotype of a heterozygote is intermediate between those of the two homozygotes, usually implying that the phenotype is more like one homozygote than the other; partial dominance.

Independent Culling   A method of selection where animals that are below a certain level of performance in one trait are not retained for breeding purposes even though they may be superior in another trait.

Interference   The decrease in the probability of a second crossover occurring near another; the closer the crossovers, the greater the interference.

Interphase   The stage in the life of a cell between mitotic divisions.

Karyotype   A picture of the chromosomes as they appear in mitotic prophase arranged in pairs according to size and position of centromeres.

Lethal Gene   A gene whose effect is drastic enough to cause the death of an individual; death may occur any time from fertilization to advanced age.

Linebreeding   A kind of inbreeding where an attempt is made to concentrate the genes of one particular ancestor into the pedigree.

Linkage   Refers to two or more genes on the same chromosome.

Linkage Group   All the genes that might exist on one chromosome.

Locus (pl., loci)   The position or location on a chromosome where a specific gene can be found.

Map Unit   A unit of measurement in linkage studies where the overall length of any chromosome is taken as 50 units, the maximum percentage of crossing-over that can occur; one-fiftieth the length of a chromosome.

Mass Selection   A kind of selection where animals under consideration for breeding purposes are compared to all others in the herd or population.

Mean   A value determined by adding several values and dividing the sum by the number of values added; average.

Median   The middle value of a series of values arrayed according to magnitude.

Meiosis   A kind of cell division associated with the production of gametes where a diploid cell undergoes a series of two divisions to produce one or more haploid cells.

Melanin   The black or brown pigment found in skin and hair cells.

Metacentric   Referring to a chromosome with a centrally located centromere.

Metaphase   That phase of cell division between prophase and anaphase where the chromosomes become aligned on the equatorial plane of the cell.

Migration   Refers to the movement of a population sample from one geographical location to another.

Mitosis   A kind of cell division associated with tissue growth and maintenance of cell numbers where a diploid cell divides to produce two identical diploid daughter cells.

Monoecious   Refers to individuals producing both sperm and ova.

Monohybrid   An individual heterozygous for one pair of genes.

Monohybrid Cross   A cross between two individuals heterozygous for one pair of genes, such as $Aa \times Aa$.

Monorchid   A male mammal with one testis in the scrotum and one inside the abdominal cavity; one of two kinds of cryptorchids.

Mutation   A chemical change in a gene resulting in the formation of an allele.

Nonadditive Genes   Genes that express themselves in a dominant or epistatic manner.

Nondisjunction   The failure of homologous chromosomes to separate during anaphase I of meiosis.

Oocyte   Any of several intermediate cells in the process of oogenesis; ovicyte.

Oogenesis   Ovigenesis; the process of producing the female gamete.

Oogonium   Ovigonium; the first or primary germ cell from which the ovum develops.

Outbreeding   A mating system involving individuals who are less closely related than the average of the population.

Outcrossing   A form of outbreeding involving relatively unrelated animals from the same breed.

Overdominance   A kind of interaction between alleles such that the heterozygote is superior to either homozygote.

Ovicyte   *See* Oocyte.

Ovigenesis   *See* Oogenesis.

Ovigonium   *See* Oogonium.

Ovine   Refers to sheep.

Oviparous   The production of progeny from eggs that hatch outside the body of the dam.

Oviposition   In birds, the process of laying an egg.

Ovoviparous   Refers to animals that produce eggs which are incubated inside the body and hatch outside the body a short time after oviposition.

Ovulation   The process of releasing eggs or ova from ovarian follicles.

Ovum   The female gamete or reproductive cell.

P₁ Cross   A parental cross, usually involving individuals homozygous for different alleles, such as $AA \times aa$ and $AABB \times aabb$.

Paint   A color pattern in horses involving white patches on a dark background.

Palomino   A copper, blond, or golden color in horses, usually with a lighter mane and tail.

Parthenogenesis   A kind of reproduction where an egg or ovum develops into an organism without having been fertilized.

Partial Dominance   A kind of allelic interaction where the phenotype of the heterozygote is intermediate between those of the homozygotes, but more similar to one homozygote than the other; similar to incomplete dominance.

Pedigree   A list of an animal's ancestors.

Penetrance   The proportion of time a gene or genotype is expressed compared to how often it is expected.

Phenotype   The way an animal appears, performs, or behaves with respect to one or more traits.

Piebald   A color pattern in mammals involving white spots on a darker background; sometimes referring to white spots on only black.

Pinto   A color pattern in horses involving white spots on a dark background.

Pleiotropy   A situation where one gene affects two or more traits.

Polygenes   Two or more (usually many, and with cumulative effects) pairs of alleles affecting a single trait such as growth rate and milk production; some alleles contribute to the trait, called contributing or additive genes, while other alleles do not.

Polyploidy   Having more than two sets of chromosomes, such as triploidy ($3n$) and tetraploidy ($4n$).

Polyspermy   The entrance of more than one sperm cell into an ovum at fertilization.

Population   A group involving all the individuals in a breed, species, geographical region, or other category, usually involving infinite numbers.

Porcine   Refers to swine.

Prepotent   The ability of an animal to transmit its characteristics to its progeny with a high degree of success.

Probability   The likelihood or chance of occurrence of an event.

Progeny   The offspring of an animal.

Progeny Testing   Determining the breeding value of an animal by studying the performance of the progeny.

Prophase   The first phase of cell division wherein many preparatory events occur, such as shortening and thickening of the chromosomes, division of the centromere, disappearance of the nuclear membrane, and formation of the spindle.

Purebred   An animal produced from several generations of breeding and selection among animals within a breed or line such that they become more homozygous than animals that are not purebred; usually associated with at least mild inbreeding.

Qualitative Traits   Traits determined by a few genes with clear distinctions among phenotypes.

Quantitative Traits   Traits determined by many genes (polygenes) with no sharp distinctions among phenotypes.

Random Mating   A mating system with no selection involved, where each male has an equal chance of mating with each female.

Reach   *See* Selection Differential.

Recessive   Refers to a gene whose expression can be modified or suppressed by the dominant allele.

Reciprocal Cross   A second cross involving the same genotypes of a first cross, but where the sexes are reversed.

Recombination   The formation of genotypes resulting from the various possible combinations of gametes of the parents.

Reduction Division   Another term for the first meiotic division wherein chromosome number is reduced.

Relationship   Genetically, the proportion of genes that two individuals possess which are the same.

Relationship Coefficient   A quantitative measure of the proportion of genes which two animals have in common because they are members of the same family.

Repeatability   A measure of the similarity between or among observations of the same trait taken at different times in the life of an animal.

Repulsion   Refers to a linkage pattern where dominant genes are linked with recessive genes.

Roan   A color pattern in cattle and horses produced by the interspersion of white hairs among hairs of a contrasting color.

Sample   A group of individuals chosen from a larger population, usually of limited numbers and intended to be representative of the larger population.

Scurs   A horny substance that grows from the skin at the location of the horn pits in polled animals, and has no bony core.

Selection   A process that causes or allows certain animals to mate and produce progeny, and prevents others from doing so.

Selection Differential   The difference between the average of a selected population or sample and the average of the population from which they were selected.

Selection Index   A measure of the average merit of animals for two or more traits.

Selection Intensity   The ratio of the selection differential to the standard deviation of a trait.

Sex Chromosomes   A pair of chromosomes in animals that determine the sex of progeny depending on which ones they obtain; one sex usually has two of one kind, the other sex having two different kinds; in mammals the male is XY and the female is XX; in birds, males are ZZ and females are ZW.

Sex-Influenced Traits   Traits affected by autosomal genes showing dominance and recessiveness between alleles; however, the allele which is dominant in one sex will be recessive in the other. Dominance and recessiveness is determined by the presence of male and female sex hormones.

Sex-Limited Traits   Traits that are expressed by only one sex although the genes are carried by both sexes.

Sex-Linked Genes   Genes carried on the sex chromosomes.

Sib Brother or sister.

Sire The male parent.

Somatic Refers to body cells or tissue as opposed to sex cells.

Sorrel A yellowish-red color in horses; lighter shades of chestnut.

Species A group of animals with certain common characteristics that when mated together produce fertile offspring.

Sperm A spermatozoon, a male sex cell.

Spermatid A haploid cell produced from the second meiotic division of spermatogenesis that has not undergone the changes to form a sperm cell.

Spermatogenesis The process in male animals of producing spermatozoa.

Spermatogonium A primary germ cell located in the testes that will undergo spermatogenesis to form spermatozoa.

Spindle The array of what appears to be fibers extending from one centriole to the other in a cell in prophase or metaphase.

Stallion A mature uncastrated male horse.

Standard Deviation A statistical measure of variation in a sample or population; the square root of the variance.

Synapsis The pairing or coming together of doubled homologous chromosomes during prophase I of meiosis.

Telocentric Refers to chromosomes with a terminal centromere.

Telophase The phase of cell division between anaphase and complete separation into daughter cells; includes the reformation of the nuclear membrane and return of the chromosomes to relatively long, slender, threadlike structures.

Testcross A genetic cross where an individual expressing a dominant trait is mated to one expressing the recessive trait, one purpose being to determine whether the dominant-bearing individual is heterozygous.

Tetrad A pair of synapsed homologous chromosomes observed during prophase I of meiosis.

Trait Any observable feature or characteristic of an animal.

Trait Ratio Ratio of the value of a trait in an individual to the mean of the herd or population.

Transgressive Variation Refers to the appearance of progeny whose phenotypes are more extreme than those of the parents or grandparents.

Trihybrid An individual heterozygous for three pairs of genes.

Trinomial A mathematical expression consisting of three terms or values.

Triploidy A condition where a cell possesses three sets of chromosomes; $3n$.

Variance A statistical measure of variation in a sample or population; the square of the standard deviation.

Viviparous The bringing forth of live offspring from the body as in mammals, as opposed to the laying and hatching of eggs.

Zona Pellucida The relatively thick covering or membrane forming the outer surface of a mammalian ovum.

Zygote A diploid cell formed by the union of a sperm cell and an ovum.

Solutions to Even-Numbered Problems

CHAPTER 1

1.2.

(a)   (b)   (c)

(d)   (e)   (f)

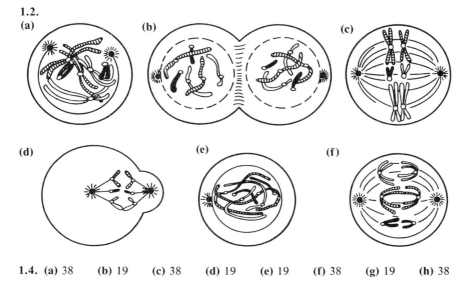

1.4. **(a)** 38   **(b)** 19   **(c)** 38   **(d)** 19   **(e)** 19   **(f)** 38   **(g)** 19   **(h)** 38

CHAPTER 2

2.2. The gene for white feathers within this group of chickens exists in the homozygous state.

2.4. (a) codominance   (b) cream

2.6. (a) All MN   (b) No. A child of group N would have to have obtained an N gene from his father; this man does not carry this gene.

2.8. No. Palomino is due to heterozygosity, which will not "breed true."

2.10. (a) $\frac{1}{2}$ hairless : $\frac{1}{2}$ normal   (b) $\frac{2}{3}$ hairless : $\frac{1}{3}$ normal

CHAPTER 3

3.2. (a) 8   (b) 4   (c) 2   (d) 1   (e) 64   (f) 4

3.4. 9

3.6. (a) 1 *BBPp* black, pea comb
          1 *BBpp* black, single comb
          1 *BbPp* blue, pea comb
          1 *Bbpp* blue, single comb
     (c) 1 *BBPp* black, pea comb
          2 *BbPp* blue, pea comb
          1 *bbPp* white, pea comb
          1 *BBpp* black, single comb
          2 *Bbpp* blue, single comb
          1 *bbpp* white, single comb

(b) 1 *BbPp* blue, pea comb
     1 *Bbpp* blue, single comb
     1 *bbPp* white, pea comb
     1 *bbpp* white, single comb
(d) All *BbPp* blue, pea comb

(e) 1 *bbPP*
     2 *bbPp*   3 white, pea comb
     1 *bbpp*   1 white, single comb

3.8. (a) (3 short-haired : 1 long-haired) × (2 Pelgers : 1 normal) = 6 short-haired Pelgers : 3 short-haired normals : 2 long-haired Pelgers : 1 long-haired normal

(b) No. Only the cross *PP* × *pp* would always produce Pelgers only, and the *PP* genotype would not live to reproduce.

(c) 3   (d) 4

CHAPTER 4

4.2. 3: TTH, THT, HTT

4.4. $\frac{1}{6} = 0.167 = 16.7\%$

4.6. 6: 1 + 6, 2 + 5, 3 + 4, 4 + 3, 5 + 2, 6 + 1

4.8. $\frac{1}{52} = 0.0192 = 1.92\%$

4.10. $(\frac{2}{13}) (\frac{4}{51}) = 8/663 = 0.012$, or 1 chance in 83

4.12. $\frac{1}{8} = 0.125 = 12.5\%$

4.14. $(\frac{9}{16})^2 = 81/256 = 0.316 = 31.6\%$

4.16. $(\frac{3}{4})^5 (\frac{7}{8})^6 = (0.237) (0.449) = 0.106 = 10.6\%$ (see Table 4–1)

4.18. (a) $5p^4q = 5 (\frac{1}{2})^4(\frac{1}{2}) = 5 (\frac{1}{2})^5 = 5/32 = 0.15625 = 15.625\%$

(b) $10p^3q^2 = 10(\frac{1}{2})^3(\frac{1}{2})^2 = 10(\frac{1}{2})^5 = 10/32 = 0.3125 = 31.25\%$

(c) $10p^2q^2 = 10(\frac{1}{2})^2(\frac{1}{2})^3 = 10(\frac{1}{2})^5 = 10/32 = 0.3125 = 31.25\%$

(d) $5pq^4 = 5(\frac{1}{2})(\frac{1}{4})^4 = 5(\frac{1}{2})^5 = 5/32 = 0.15625 = 15.625\%$

(e) $p^5 = (\frac{1}{2})^5 = 1/32 = 0.03125 = 3.125\%$

4.20. (a) 2   (b) 1   (c) 3   (d) 3   (e) 4

4.22. $\chi^2 = 0.20$, P is less than 0.70

4.24. Compared to the expected ratio of 7 : 35 : 70 : 70 : 35 : 7, the $\chi^2 = 5.12$, df $= 5$, probability between 0.50 and 0.30.

CHAPTER 5

5.2. (a) Among the female kittens the expected colors and proportions are $\frac{1}{4}$ solid black, $\frac{1}{4}$ black with white spots, $\frac{1}{4}$ tortoiseshell, $\frac{1}{4}$ calico; among the male kittens the expectations would be $\frac{1}{4}$ solid black, $\frac{1}{4}$ spotted black, $\frac{1}{4}$ solid yellow, $\frac{1}{4}$ spotted yellow   (b) probably solid yellow but carrying a spotting gene

5.4. (a) $\frac{1}{3}$ barred creeper males, $\frac{1}{3}$ nonbarred creeper females, $\frac{1}{6}$ barred normal males, $\frac{1}{6}$ nonbarred normal females   (b) homozygous barred creeper

5.6. (a) zero   (b) All females would obtain the recessive gene from their sire and be gold colored; one-half would be heterozygous for feathering types, one-half would be homozygous recessive, but all would be hen feathered.   (c) All chicks will develop gold plumage, $\frac{3}{4}$ of the males would be expected to become hen feathered, $\frac{1}{4}$ of the males should develop cock feathering, all females should be hen feathered.

CHAPTER 6

6.2. (a) Either $\frac{3}{4}$ mallard to $\frac{1}{4}$ restricted, or $\frac{3}{4}$ restricted to $\frac{1}{4}$ mallard depending on which of the two genes involved is dominant.   (b) The expected genotypic ratio is $\frac{1}{3}$ *Mm,* $\frac{1}{6}$ *MM,* $\frac{1}{6}$ *Md,* $\frac{1}{6}$ *md,* $\frac{1}{6}$ *mm;* the expected phenotypic ratio is $\frac{2}{3}$ mallard, and $\frac{1}{3}$ restricted mallard.

6.4. (a) $\frac{1}{4}$ (b) Approximately $\frac{3}{8}$ of the children should be type AMN, $\frac{3}{8}$ type AN, $\frac{1}{8}$ OMN, and $\frac{1}{8}$ ON.

CHAPTER 7

7.2. (a) 0.725   (b) 52.6% chestnut, 39.8%, palomino, 7.6% white   (c) The breeder appears to be favoring palomino.

7.4. (a) 0.0253   (b) five   (c) approximately 1500

7.6. (a) 0.13   (b) 53   (c) 24   (d) Yes. Nonscurred bulls mated to scurred cows or heterozygous nonscurred cows would be expected to produce some bull calves that would develop scurs, but no scurred heifers.

7.8. (a) 0.107   (b) 4 yellow, 65 tortoiseshell, 269 nonyellow   (c) 38 yellow, 315 nonyellow   (d) Observed numbers are close enough to expected numbers to assume that random mating had occurred.

CHAPTER 8

8.2. (a) $\frac{1}{2}$ black (*BbCc* and *Bbcc*), $\frac{1}{4}$ red (*bbCc*), $\frac{1}{4}$ yellow (*bbcc*)   (b) $\frac{3}{4}$ black, $\frac{3}{16}$ red, $\frac{1}{16}$ yellow

8.4. (a) 0.25 (probability between 50 and 75%)   **(b)** 0.5 (probability between 40 and 50%)   **(c)** approximately 246

CHAPTER 9

9.2. (a) yes   **(b)** 21.2   **(c)** *FFcc* × *ffCC*   **(d)** 1%

9.4. (a) oval cells with Rh positive, and normal cells with Rh negative   **(b)** 0.10   **(c)** 50%   **(d)** 80%

CHAPTER 10

10.2 (a) 16 mm and 40 mm   **(b)** *aabbcc* and *AABBCC*   **(c)** 7   **(d)** 64   **(e)** 1 with 16 mm, 6 with 20 mm, 15 with 24 mm, 20 with 28 mm, 15 with 32 mm, 6 with 36 mm, and 1 with 40 mm   **(f)** transgressive variation   **(g)** 4 pairs

CHAPTER 11

11.2. (a) 79   **(b)** 26.8   **(c)** 5.2   **(d)** 0.066

11.4. (a) 2.15   **(b)** 1.91   **(c)** between 5 and 10%

11.6. (a) 124.5   **(b)** 30.0   **(c)** −43.0   **(d)** −0.35   **(e)** 8.24   **(f)** −0.704

CHAPTER 12

12.2. (a) 8.5 mm   **(b)** −3.8 mm   **(c)** 34.2 mm

12.4. (a) 1600 lb²   **(b)** 480 lb²   **(c)** 608 lb²   **(d)** 38%

12.6. (a) 42.4%   **(b)** 41.3%   **(c)** 51.9%

CHAPTER 13

13.2. (a) 1.2 lb   **(b)** 1.9 lb   **(c)** 8.46 lb   **(d)** 8.18 lb   **(e)** 8.39 lb

13.4. (a) sow numbers 4 (243), 9 (237), 5 (236), 2 (232), and 7 (230)   **(b)** 23.6 units   **(c)** 225

13.6 (a) 906 lb   **(b)** 863 lb   **(c)** 887.5 lb

CHAPTER 14

14.2 0.03125

$(\frac{1}{2})(\frac{1}{2})^4 = 1/32 = 0.03125$

14.4. 0.1216

(6 pathways)

$$\frac{[(\tfrac{1}{2})^3(1.03125)] + [(\tfrac{1}{2})^4(1.03125)] + (\tfrac{1}{2})^5 + (\tfrac{1}{2})^6 + (\tfrac{1}{2})^9 + (\tfrac{1}{2})^{10}}{2} = 0.1216$$

14.6. $0.4068;$ $\dfrac{\tfrac{1}{4} + \tfrac{1}{8} + \tfrac{1}{16}}{\sqrt{(1.03125)(1.1216)}} = \dfrac{0.43750}{1.07548} = 0.4068$

14.8. 0.3513

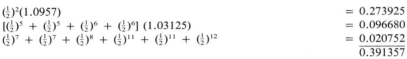

$$
\begin{aligned}
(\tfrac{1}{2})^2(1.0957) &= 0.273925 \\
[(\tfrac{1}{2})^5 + (\tfrac{1}{2})^5 + (\tfrac{1}{2})^6 + (\tfrac{1}{2})^6] (1.03125) &= 0.096680 \\
(\tfrac{1}{2})^7 + (\tfrac{1}{2})^7 + (\tfrac{1}{2})^8 + (\tfrac{1}{2})^{11} + (\tfrac{1}{2})^{11} + (\tfrac{1}{2})^{12} &= \underline{0.020752} \\
& 0.391357
\end{aligned}
$$

$$R = \frac{0.391357}{\sqrt{(1.1064)(1.1216)}} = \frac{0.391357}{1.113974} = 0.3513$$

CHAPTER 15

15.2. 3.80%

15.4. (a) 6.0% (b) 11.6% (c) 18.4%

Index